Building Foundations in Science Series

Fundamentals of Physical Science
Physics

Michael J. Anzelone, BS, MS

Editor:
Erin Conklin-Kimaoui, BA

Acknowledgements:
This book is dedicated to my wife Rita and daughters Danielle and Michelle.
Special thanks to Michelle Smedley, West Islip, NY Public Schools for her comments regarding mathematics concepts.
Special thanks also to Fred L. Wehran, BSCE, MSCE., Consulting Engineer, Mahwah, NJ.

Digital Composition and Line Art:
Michael J. Anzelone

Cover design by Danielle Anzelone

© 2012 This book is protected by copyright. All rights reserved.

No part of this publication may be reproduced or distributed in any form: photocopy, electronic, mechanical, recording or by any other means, or stored in a database or retrieval system, or transmitted without the prior written permission of the publisher.

Published by Angelus 1, Inc.
PO Box 64
Williston Park, NY 11596-0064

Table of Contents

To the Student..6
General Laboratory Safety Rules... 7
Chapter 1 The Development of Scientific Thinking................................. 8
    Science and Technology...8
    Science and Technology in Ancient Civilizations......................................8
    Medieval European Natural Philosophy (The Dark Ages)..................13
    Pre-Renaissance Italy, Early to Later Middle Ages..............................14
    Early Renaissance and the Rise of Modern Science.............................15
    The Age of Discovery and Exploration...17
    Mid-Renaissance...18
    Late Renaissance Leads to the Age of Enlightenment........................19
    The Enlightenment...20
    Technology and Invention in the 1800s...22
    The Second World War Spurs on Scientific Achievement..................28

Chapter 2 Science and the Scientific Method...31
    The Natural Sciences...31
    The Biological Sciences...32
    The Physical Sciences..34
    The Scientific Method..35
    Forms of Trustworthy Scientific Information.......................................38
    Forms of Information That Are Not Trustworthy................................38
    Making False Claims by Pretending to be Scientific............................40

Chapter 3 Measurement..41
    Measurement Defined..41
    Length..44
    Volume...44
    Mass..45
    Estimation...46
    Precision and Accuracy..46
    Additional Units of the of SI System of Measurement.......................47
    Kelvin, Celsius and Fahrenheit Temperature Scales...........................48
    Significant Digits..49
    Scientific Notation..49

Chapter 4 Introduction to The Science of Physics..................................50
    Basic Concepts in Physics...50
        Matter..50
        Mass...50
        Length...50
        Area..51
        Volume..51
        Units and Standards...51
        Ratio and Division...51
        Vertical and Horizontal...51

　　　　Kinetic Theory of Matter..52
　　　　Gravity...52
　　　　Weight..52
　　　　Writing and Solving Algebraic Equations........................52
　　　　Density...53
　　　　Specific Gravity..53
　　　　Energy.. 53
　　　　Forms of Energy...53
　　　　Machines..54
　　　　Friction...54
　　　　Force...54
　　　　Measuring Forces..55
　　　　Scalars...55
　　　　Graphing Data..55

Chapter 5 Motion..57
　　　　One-Dimensional Kinematics:Motion in One Direction...........57
　　　　Acceleration...60
　　　　Forces Act on Bodies..61
　　　　Newton's Laws of Motion...62
　　　　Two-Dimensional Kinematics: Motion in Two Directions............66
　　　　Motion in Circles...68
　　　　Law of Conservation of Momentum............................. ...69

Chapter 6 Energy..70
　　　　Kinetic Energy 70
　　　　Vertical Distance and Horizontal Distance as They Apply to Work.71
　　　　Conservation of Energy..71
　　　　Temperature...72
　　　　Thermal Energy and Heat...72

Chapter 7 Machines..73
　　　　　　The Lever..74

Chapter 8 Solids, Liquids and Gases...78
　　　　The Structure of Matter.. 78
　　　　States of Matter...79
　　　　Kinetic Theory of Matter..80
　　　　Pressure..82
　　　　Boyle's Law...82
　　　　Charles' Law..83
　　　　Buoyancy..84
　　　　Archimedes' Principle...84
　　　　Pascal's Principle...85
　　　　Bernoulli's Principle..86
　　　　Change of State..87

Chapter 9 Waves and Sound...90
　　　　Characteristics of a Wave...90
　　　　The Nature of Sound...90

The Doppler Effect..92
Practical Application of the Uses of Sound Waves.......................93

Chapter 10 Light..**94**
    Transverse Waves...94
    The Electromagnetic Spectrum...95
    Wave Properties of Light...100
        Transmission...101
        Reflection..101
        Refraction...101
        Diffraction..102

Chapter 11 Mirrors and Lenses..**104**
    Plane Mirrors..104
    Curved Mirrors...104
    Lenses...105
    Vision..106
    Applications of Light Waves..107
    Polarized Light...110

Chapter 12 Electricity..**111**
    Static Electricity...111
    Conductors..111
    Insulators..112
    Electric Current..113
    Resistance and Ohm's Law..113
    Electrical Circuits...114
    Electric Power.. 116

Chapter 13 Magnetism..**117**
    Magnets..117
    Magnetic Fields..118
    Effect of Electric Current...119
    Direct Current...122
    Alternating Current..123
    Generators..123
    Motors..123
    Transformers..124
    Electronics..125

Chapter 14 Radioactivity..**126**
    Radioactive Elements and Radioactive Decay....................126
    Nuclides..127
    Half-Life and Radioactive Decay..128
    Carbon-14 Dating...129
    Nuclear Fission...129
    Nuclear Fusion... 130
    Detecting Radioactivity and Radiation Counters................131
    Unified Physics.. 132

TO THE STUDENT

This text, if used conscientiously, provides a foundation in the physical sciences that would ordinarily require years of study. The basic concepts presented in this text are essential for successful completion of more advanced courses in physics and health-related sciences such as nursing. The concepts presented in this text are simple, yet complete enough for the beginning student.

Year after year I hear students say "they (meaning other students) get good grades because they are smart." It only seems that way. It is not magic. These "smart" students bring a good foundation in science to their new studies. Most have taken science in elementary school through high school. In the course of their education, many of the same ideas are repeated every year. This builds a good foundation. A good foundation makes learning and integrating new information easier. Those who are beginning their study without a good background will have a more difficult time assimilating new knowledge, but it can be done. The solution is additional reading.

This text is different from most general science review books. It is a historically, and culturally-based presentation of science. Science has a rich and interesting history that took place in a cultural context and still does today. Science does not exist in a vacuum. The practice of science is part and parcel of a culture. An understanding of the roots of science and scientific discovery will allow scientific concepts to be remembered easily.

Repetition is liberally employed in this text. As concepts are explored in depth, elementary concepts are reviewed. Elementary school concepts through college levels are bridged. The flow of information is regulated to build a knowledge base for the new or experienced student.

This text should be read in a thorough manner. Pay attention to headings and subheadings. They are the introductions to an explanation of concepts that follow. Keep these headings in your mind while reading.

There are over one hundred explanatory annotations in the text. Refer to them. Too many students do not turn to a glossary or use a dictionary when they do not understand a word or an idea in a text. I have placed the equivalent of both in the form of explanatory annotations at the bottom of each page. They will serve the student who has been out of school for an extended period and those for whom English is a second language. I have tried to bridge both gaps. In addition, simple illustrations of important concepts are a vital feature of this text. Refer to them. Illustrations will clarify the reading. Read the legends that appear under or next to a diagram or photograph.

Legge Felicitas
Michael J. Anzelone

GENERAL LABORATORY SAFETY RULES

1. *The most important safety rule is to listen and read directions before touching any equipment or starting an exercise.* If there is any spare time in lecture or laboratory, READ – NEVER WASTE TIME.

2. Know the location of fire extinguishers, fire blankets, eyewash, and fire showers and how they are used.

3. No eating in the laboratory.

4. Keep the work area on the bench *neat*, *clean* and *organized*.
    a. Coats, books, and other materials not needed for lab must be placed in a locker.
    b. Never place coats or any other clothing on the backs of chairs.

5. Handle everything as if it were dangerous, or in the case of microbiology-pathogenic.

6. Always disinfect a work area if biohazards are used. Wash hands thoroughly before and after working.

7. Make sure no one is close by when transporting anything that is *hot*, *sharp* or *caustic*.

8. *Notify the instructor of anything that breaks or spills.* Students should not pick up broken glass or spilled materials. Let the instructor or custodians do it.

9. Use proper protection for a given situation.
    a. If something can splash or break and fly off, use goggles and wear lab coats.
    b. Wear gloves or use tongs when handling anything hot or potentially harmful.
    c. Long hair should be tied back when working with a flame.
    d. Loose clothing should be removed or secured in some manner.

10. A student should never work alone in the laboratory.

11. Keep flames far away from flammable solutions.

12. Clean glassware before use, even if it looks clean.

13. Clean glassware before putting it back to where it belongs.

14. The person that took material from stock should be the person to replace it to stock.

15. When weighing out chemicals or pouring reagents, do not put the excess material back into the stock bottle.

16. Any needles or scalpels must go into a sharps box.

17. Never force any piece of equipment beyond what it is supposed to do.

Chapter 1 The Development of Scientific Thinking
Science and Technology
Technology is defined as using knowledge for some useful purpose. Many of humankind's most important early discoveries were technological in nature. Ancient man knew how *to do* things. They did not have an *understanding* of the underlying processes. Science began with humankind's use reason to understand the cause of events.

Ancient man in Africa used fire for practical purposes. They used fire to keep warm, cook food and keep away predators. They did not know it was an exothermic chemical reaction that gave off heat and light.

Science and Technology in Ancient Civilizations
Africa (c[1]. 100,000 B.C. – 50,000 B.C.)

Fossil[2] evidence indicates the earliest humans, ***Homo***[3] ***sapiens***,[4] originated in Africa about 100,000 to 200,000 years ago. The most important discoveries made by these ancient humans are the use of tools, fire and numbers. Some humans left Africa and migrated to India, Europe, Asia and China around 50,000 to 60,000 B.C.

China (6000 B.C. – 1279 A.D.)

The ancient Chinese civilization developed culture, literature and philosophy c. 1045-256 B.C. The four great technological inventions of the Chinese people were *papermaking*, *the compass*, *gunpowder* and *printing*. They were the first to publish a written Canon of Medicine and establish a university c. 124 B.C.

Mesopotamia[5] (c. 5300 B.C. – 600 A.D.)

Peoples of Mesopotamia invented writing (3500-3000 B.C.), made tools and had libraries. They had a concept of **physical examination**,[6] **diagnosis**,[7] **prognosis**[8] and **prescription**.[9] However, if illness could not be cured, then **exorcism**[10] was employed. Babylonians had a true place-value system.

---

1   Around or about.
2   A preserved part or an imprint of a living thing in stone or some other medium.
3   Latin for man.
4   Latin for wise or wisdom.
5   Mesopotamia is located between the Tigris and Euphrates rivers. Primarily what is present day Iraq, and parts of Syria, Turkey and Iran.
6   Performed after a patient's history is taken. During the examination the patient is examined for signs of disease. Signs are *objective* such as temperature or a growth on the body. Signs are responsible for the symptoms a patient may feel. Symptoms are non-measurable *subjective* feelings noticed by a patient such as malaise, chills or nausea.
7   Discovering a cause of a disease from a patient's history, signs and symptoms.
8   Determining the possible outcome of a disease.
9   Written or oral direction that directs a plan of care for a patient, a "care plan."
10   A ritual that will drive an evil spirit from an ill individual.

India: *(3000 B.C. – 1190 A.D.)*
Ancient Indian civilization developed irrigation, sewers, **reservoirs**,[11] cultivation of cotton and sugarcane, indoor bathrooms connected to drainage systems, horse-drawn plows, astronomical texts (c. 300 B.C.), cataract surgery, the concept of 0 and inoculation against smallpox (c. 700 B.C.). Indian mathematicians invented what we know today as "Arabic numerals" (0, 1, 2, 3, 4, 5, ...). The Hindu-Arabic number system eventually made its way to Europe.

Greece: *(750 – 146 B.C.)*
The ancient Greeks developed the concept of science. They were the first to attempt to organize the world around them. *Science, to the ancient Greeks, was an organized seeking of how things happen.* They used observation and **philosophical argument**[12] to understand how things happen in the natural world. This was a form of scientific thinking. Some Greek scholars looked for rational explanations for the cause of things. A few ignored the supernatural completely, and believed in *natural causes for every event*. This was the beginning of scientific thinking, but it was not experimental science.

Around 400 B.C., Pythagoras (c. 570-495 B.C.) was looking for order in mathematics and came up with the idea of a **mathematical proof**.[13] Socrates (c. 469 B.C.-399 B.C.) developed methods to determine the truth or falsity of a **premise**.[14] Aristotle (384 B.C.-322 B.C.) devised systems of classifying plants and animals. Anaxagoras (c. 500 B.C.-428 B.C.) was a philosopher that tried to explain eclipses, meteors, rainbows, and the Sun in a scientific manner. He believed the mind was an instrument for ordering things. He hypothesized that all matter was composed of a number of indestructible elemental particles. Hippocrates of Cos (c. 460 B.C.-c. 370 B.C.) was an ancient Greek physician who had an impact on Western medicine which lasted for thousands of years. He is considered to be the "Father of Medicine." See table 1.1.

For the most part, the ancient Greeks arrived at scientific truths by argument. Their methods were **philosophical**[15] in nature, not experimental, but the ancient Greeks planted the seeds for modern science that grew and bloomed in Italy between the later Middle Ages and the height of the Italian Renaissance.

11　French meaning storehouse.
12　To use logical thinking to prove a point, to persuade or make something clear.
13　A logical mathematical statement arrived at by deductive reasoning.
14　A premise is a statement that is assumed to be true.
15　Philosophy. From the Greek – *philo* meaning love and *–sophos* meaning wisdom.

Table 1.1. Greek thinkers build the foundation for Western science.

Philosopher	Achievement(s)	Philosopher	Achievement(s)
Thales (c. 640-546 B.C.)	First of many to believe in cause and effect.	Plato (c. 427- c. 347 B.C.)	Founded the Academy in Athens (first institution of higher learning in the Western world).
Alcmaeon of Croton* (500 B.C.) *Town in southern Italy.	First to dissect humans for research. He described internal structures, nerves of the eye (anatomy), brain as center of intelligence, development of chick in an egg (embryology).	Hippocrates (460 B.C.- 375 B.C.)	**Father of Medicine.** Major influence on medical treatment to this day.
Democritus (460-370 B.C.)	Atomic theory or atomism (resembles modern atomic theory), biological investigations too.	Aristotle (384-322 B.C.)	**The First Biologist.** Looked for causes of phenomena, first to make distinctions between different disciplines of biology, organic diversity, classification of invertebrates.
Socrates (469-399 B.C.)	A founder of Western philosophy. He, Plato and Aristotle helped lay the foundation of Western philosophy and science.	Archimedes of Syracuse* (287-212 B.C.) *Sicily	We would recognize his model of physics today. His treatment of buoyancy, density ($D = m/V$)†, the lever and screw pump. He used experiment to test hypotheses. †$D=density$, $m=mass$ and $V=volume$.

Ancient Italy: Roman Period (c. 700 B.C. – c. 500 A.D.)

Rome dominated Europe and countries bordering the Mediterranean for more than 1,000 years. An Italic people, the **Romans**[16] accepted many different peoples into their Republic from whom they *learned and borrowed useful ideas*. Latin was spoken as the native language in throughout the Roman world (700 B.C.-500 A.D.).

The Romans founded a great number of cities. A few among the hundreds were: London (Londinium), Cologne (Colonia Agrippina), Paris (Lutetia Parisiorum). The Romans administered an area that stretched from England in the North to the Sahara desert in the South and from

---
16   Romans were the Italic peoples living in Rome and the Italian peninsula.

Portugal in the West to Romania in the East and Syria in the South East. They brought their culture and language to these areas. *Latin became the universal language of commerce, communication and science.*

The main requirements for Roman citizenship in the monarchy, republic and the Empire were: service in the army, paying taxes and s*peaking, reading and writing* **Latin**.[17] The long period of "Romanization" of many different peoples produced a great influence on many modern languages such as Italian, Portuguese, French, Romanian, Spanish and English,

England was a Roman province for over 400 years, thereby infusing the Latin language and alphabet into the English language. Latin became the language of the educated and a universal language of communication in all Roman provinces. *All schools taught lessons in Latin. Names of organisms and anatomical structures were in Latin. This is why names of organisms and anatomical structures are in Latin in textbooks.*

The Romans did more than influence the development of language and the concept of a republic. They made great advances in engineering, technology and medicine. See table 1.2. Their invention of concrete enabled Roman engineers to build great public works such as the Pantheon, the largest unreinforced dome in the world to this day, Trajan's Market in Rome, the first shopping mall and the Colosseum, the first superdome. Their engineers constructed over 50,000 miles of civilian and military roads to all parts of the Empire. Aqueducts, indoor plumbing and sewer systems greatly improved sanitation resulting in improved general health of citizens. One of the many major achievements of the Romans was the first *public medical service*. A modern surgeon would recognize many Roman surgical instruments. Naturalists, such as Celsus and Pliny the Elder, compiled volumes of information that were useful for the study of science.

The central Roman administration gradually gave up control of public administration, the army and maintenance of public roads and buildings. The light of the classical period was dimming, and the Medieval period or The "**Dark Ages**"[18] had begun. The Latin language, however, remained as the language of the educated and was used in science, literature, law and administration.

---

17 The Latin alphabet originated in Italy c. 600 B.C. and is still in use today.
18 Also referred to as the Middle Ages or the Medieval Period. Francesco Petrarca (Petrarch 1304-1374), an Italian scholar, was one of the first to use the term "Dark Ages."

Table 1.2. Roman science, medicine and technology.

Philosopher	Achievement(s)	Philosopher	Achievement(s)
Lucretius (c. 98-33 B.C.)	*De rerum natura (On the Nature of Things).*	Claudius Ptolemy (c. A.D. 90-168)	Known for "geocentric theory of the universe." The sun revolves around the Earth.
Cicero (106-43 B.C.)	In his treatise "On Duties," a moral duty exists that all people should receive justice because they are humans.	Galen (A.D 129-200).	Great contributions to anatomy, pathology, pharmacology. His studies were based on apes. His works were the authority for over 1,000 years.
Aulus Cornelius Celsus (c. A.D. 1)	Listed four major signs of inflammation: heat, pain, redness, and swelling. Used antiseptic substances on wounds.	Dioscorides (A.D. 40-90)	Founder of pharmacology. Described medicinal properties of over 600 plants.

Islam and Islamic Science

The Romans traded with peoples of the Middle East, enabling the Arab world to come into contact with the literature and scientific works of ancient writers of Greece and Rome. During the Islamic Golden Age (A.D. c.750-A.D.-c.1258) Indian, Persian (modern Iran), Latin and especially Greek knowledge was translated into Arabic.

Islamic scientists and philosophers included Persian, **Arabs**,[19] **Moors**,[20] Turks and Egyptians. Islamic scientists were of different religious backgrounds. They were mostly Muslims, but also some Christians and Jews. Islamic scholars translated Greek and Latin texts into Arabic, thus preserving them for the future. Classical works would be sought after by early Renaissance scholars in Europe and the Islamic world. Islamic scholars did more than preserve Western knowledge. See table 1.3. They had their own "scientific revolution" (A.D. 600-A.D. 1400). Mathematics, astronomy, medicine, ophthalmology, physics, **alchemy**[21] and chemistry were some fields of study in the Islamic world.

19   Arabs are peoples of Western Asia and North Africa. Most are Muslim.
20   Moors were peoples of Western North Africa. Most were Muslim.
21   The aim of alchemy was to turn base metals into gold, find a universal remedy to cure all diseases (a panacea) and a universal solvent that would be useful in medicine.

Table 1.3. The Arab world's contributions to science and medicine.

Philosopher	Achievement(s)	Philosopher	Achievement(s)
Al-Jahiz (781-c. 868)	Described evolutionary ideas, food chain.	Abu Ali Sina (890-1037) Latinized to Avicenna (Persian–now Modern Iran)	"Canon of Medicine" and a text used in European medical schools until 1700s.
Al-Dinawari (828-896)	Wrote "Book of Plants" and described over 600 species.	Al-Hazen from Basra (965-1038)	Modern theory of optics. Light emanates from the object that is seen.
Al-Razi or-Rhazes (865-925)	Wrote the *Comprehensive Book*. Compiled Greek, Indian, and Middle Eastern medicine.	Ibn Sahl (c. 940-1,000)	Discovered the law of refraction of lenses.
Abu Kamil Shujā ibn (Auoquamel) Egyptian Muslim. (850-c. 930)	Made major contributions to algebra and geometry.	Ibn-al-Nafis (1210-1288)	The septum of the heart did not have pores to allow passage of blood.

Medieval European Natural Philosophy[22] *(the Dark Ages)*

The Medieval period (A.D. 500-1000) saw a general decline in living standards for Europeans after the year 500 because of the gradual abandoning of administration of civil and military affairs by the Romans. Infrastructure such as aqueducts for clean water, sewer systems to carry away human waste and roads fell into disrepair.

From 1095 to 1291 a period of religious warfare called the Crusades took place. Catholic Europeans battled Muslims for control of the Holy Land. The Crusades brought Europeans in contact with the Muslim world's knowledge of science, medicine and mathematics. *The knowledge of the ancient world that was preserved and used by Muslim scientists and philosophers became available to Europeans.* Much of this information was translated into Latin from the Arabic in the Early European centers of learning. Investigations were carried out in medicine, mathematics, and astronomy during this period along with the rise of medieval universities. See table 1.4.

---

22 Natural philosophy is more of a philosophical exploration as to the causes of things. Natural philosophy evolved into natural science, an experimental exploration into the cause of things. Eventually the term natural science was shortened to "science."

Pre-Renaissance Italy, Early to Later Middle Ages

The **Italian Renaissance**[23] was begun by Italian artists, engineers and philosophers of the late 1200s. They began to exhibit humanistic views of the world, a concern for human values and an interest in the classical world. The revival of human dissection and the writings of Francesco Petrarca (Petrarch) provided the impetus for change. Both heralded the emergence of a modern view of the world in *pre*-Renaissance Italy. This view evolved into a cultural movement, a ***Rinascimento***.[24] Humanism made the "scientific revolution" possible.

Revived interest in scholarship had its beginnings in the Benedictine abbey of **Monte Cassino**,[25] Italy. Established in 529, Monte Cassino's library housed the priceless ancient texts of Greek, Latin and Arabic scholars. Europe's first medical school was established here at **Salerno**,[26] c. 900, close to Monte Cassino's library. Access to this information made the medical school the most important source of medical knowledge for centuries. See table 1.4.

Table 1.4. Establishment of medical schools and universities in Europe.

Salerno, Italy (c. 900)	The *first* university to be established in Europe. It is organized around a medical school and its location is strategically located near the library at Monte Cassino, a repository for Greek and Roman knowledge.	Montpelier, France (c. 1160)	Medical school and university established.
Bologna, Italy (1088)	Second university to be established in Europe. It was organized around a law school.	Oxford, England (c. 1000s)	University
Paris, France (1150)	University	Salamanca, Spain (1218)	University
Cambridge, England (1209)	University	Padua, Italy (1222)	University

---

23   From the French–"to be reborn." Originally from the Italian–*Rinascimento*.
24   It.– a rebirth of learning.
25   The Abbey is midway between Rome and the city of Salerno, south of Naples.
26   *Schola Medica Salernitana* is still a functioning medical school.

There is a **consensus**[27] that the Renaissance began in Florence, Italy, in the late 1200s, preceded by a rebirth of learning in Italy during the late 1100s and early 1200s. The roots of the Italian Renaissance were established here. The Renaissance represented a major shift in human values that spanned about 400 years, from the 1300s to the 1700s. For the first time, large numbers of people shifted from a "God-centered" or *theocentric* way of thinking to a "man-centered" or *anthropomorphic* way of thinking. Renaissance thinking displays a greater concern for the *individual*. For this reason it is also termed a "humanist" movement that was reflected in Italian art, literature, architecture and science of the period.

There are several theories as to why the Renaissance began in Italy: (1) Italian cities were surrounded by the ruins of ancient Rome, (2) the plague, or black death, ravaged Italy first and may have caused a shift away from spirituality and caused more concern with the present, (3) a great deal of wealth was inherited by the survivors of the plague. It is probably a combination of all three phenomena.

Early Renaissance and the Rise of Modern Science in Italy
Beginning in Florence, and spreading to the rest of Italy and then to the rest of Europe by the 1700s, this 400 year period saw vast changes in art, literature, science, technology, architecture, philosophy, religion, politics and music throughout Europe. Mondino de' Luzzi, an Italian physician and professor of surgery, broke with tradition and did his own dissections to instruct his students. He wrote the first book devoted solely to anatomy in 1316.

Art and science were very much interrelated during this period. Many artists did their own dissections in order to produce a more accurate representation of the human form. *Without the Renaissance there would not have been a scientific revolution because science is a view of the world with man and man's needs at its center.*

Man is part of nature and nature began to be re-evaluated, as seen in the works of Marsilio Fecino, **Pietro Pomponazzi**[28] and many others. See table 1.5. Published in 1567,[29] Pomponazzi wrote in his *On the Causes of Natural Effects or On Incantations,* **"It is possible to justify any experience by natural causes and natural causes only."** This was a very radical thought at the time.

---

27   "Consensus" is a general agreement.
28   "When I sing a song to the Sun it is not because I expect the Sun to change its course but [because] I expect to put myself into *a different cast of mind* in relation to the Sun." Italics mine. From Bronowski, *Magic, Science, and Civilization,* 1978.
29   Published after his death.

Table 1.5. Pre-Renaissance and Early Renaissance Thinkers

Philosopher	Achievement(s)	Philosopher	Achievement(s)
Leonardo Fibonacci (1170-1240) Italian mathematician	His writings introduced Arabic numerals into European mathematics. Number theory.	Francesco Petrarca (1304-1374) Italian scholar	Considered to be the "**Father of Humanism.**"
Albertus Magnus (1193-1280) German scholar	Helped lay foundation of medieval science using rediscovered Greek and Arab knowledge.	Mondino de' Luzzi (1275-1326) Italian physician	He did his own dissections and did not defer to the authority of Galen.
Roger Bacon (c. 1220-1294) English philosopher and Fransciscan friar	A proponent of the use of experiment in the pursuit of knowledge, although he never made a scientific discovery.	Emmanuel Chrysoloras (c. 1355-1415) Turkish scholar	A major force in reintroducing Greek literature to Western Europe.
St. Thomas Aquinas (1225-1274) Italian priest and philosopher	Tried to make Aristotle's philosophical views compatible with Christian doctrine.	Johann Guttenberg (1388-1468) German goldsmith	Invented the printing press.
Giotto (c. 1266-1337) Giotto di Bondone. Italian painter and architect	One of the first great artists leading up to the Italian Renaissance. He painted the natural world as well any artist of *all time*.	Luca Signorelli (c. 1445-1523) Italian painter	Dissections of humans obtained from cemeteries. Mastery of the human form.
William of Okham (1288-1348) English scholar	"Okham's Razor" or the *principle of parsimony*. The simplest explanation is usually best.	Pietro Pomponazzi (1462-1525) Italian philosopher	"It is possible to justify any experience by natural causes and natural causes only."

The Age of Discovery and Exploration

Marco Polo's explorations and the voyages of Christopher Columbus were the most important factors that produced a surge in exploration and discovery in the 1400s and 1500s. The Polo family were merchants from Venice, Italy that had traded with the East for many years.

In 1271, 17 year old Marco Polo, his father and uncle traveled for 25 years through Asia. Upon his return to Italy in 1300, he authored the book, in 1300 we have come to know as *The Adventures of Marco Polo*. The book told of the wondrous sights and adventures he encountered in Asia.

As a result of exploration and the discovery of new lands, new species of plants and animals were discovered that were not found in Europe. The newly discovered **flora**[30] and **fauna**[31] widened the horizons of natural philosophers and scientists of the Renaissance. See table 1.6.

Table 1.6. Explorers widen the scope of living things through their discovery of new lands.

Explorer	Achievement(s)	Explorer	Achievement(s)
Marco Polo (c. 1254-1324) Italian merchant and explorer	Traveled 25 years through Asia and China. Gave a detailed view of Asia.	Henry Hudson (1560/70-1611?) English explorer	Sailed for the Dutch East India Company. Explored the Hudson River in New York.
Cristoforo Colombo (Christopher Columbus) (1451-1506) Italian mariner.	Sailed for Spain. Discovers "New World" in 1492.	Vasco da Gama (c. 1460 or 1469-1524) Portuguese explorer	First to sail from Europe directly to India.
Giovanni Caboto (John Cabot) (1450-1499) Italian explorer that sailed for England	Discovered North American continent (Newfoundland) in 1497.	Ferdinand Magellan (1480-1521) Portuguese explorer	First to circumnavigate the globe (1519-1522).
Amerigo Vespucci (1454-1612) Italian mariner. Sailed for Portugal. "America" derived from his first name.	Explored the East coast of South America. Determined North and South are separate continents.	Giovanni Verazzano (1485-1528) Italian mariner. Sailed for France.	Explored Coast of South Carolina to Newfoundland and New York harbor in 1524.

---

30   Plant life.
31   Animal life.

Mid-Renaissance
The Spark of Reason Spreads to the Rest of Europe

The foundation for the Renaissance was put down from the late 1200s through the 1300s. The Renaissance did not "just happen." Around the 1500s, a new view of the world became very noticeable.

"Something had happened in Italy that made a great inroad in established, authoritarian, and traditional views of life."[32] Philosophers, artists, architects and engineers no longer automatically deferred to authority. They looked at nature from a *human perspective*, not from a perspective they were told to have. Italian society was becoming "man"-centered, as opposed to being "God" centered. Some of the greatest artists/anatomists are seen during this period. If artists were to present a more realistic representation of the human form, then they had to become anatomists. Artists probably had a better understanding of anatomy than those trained in medicine. See table 1.7.

Table 1.7. Scientists, philosophers and artists of the Mid- to Late Renaissance.

Scientist/Artist	Achievement(s)	Scientist/Artist	Achievement(s)
Leonardo da Vinci (1452-1519) Italian artist, anatomist, engineer and inventor	Dissected humans and animals and illustrated his findings. Made great advances in many areas of science.	Matteo Realdo Colombo (Renaldus Columbus) (1516-1559) Italian professor of anatomy and surgeon	Discovered pulmonary circulation of the blood. Aided in William Harvey's discovery of human circulation of blood.
Nicolaus Copernicus (1473-1543) Polish astronomer	Heliocentric theory of the universe. The Earth revolves around the Sun.	Johannes Kepler (1594-1600) German astronomer	Laws of planetary motion.
Girolamo Fracastoro (1478-1553) Italian physician and scholar	Epidemic diseases caused by tiny particles that transmit infection. Coined the word "fomite."	Gabriele Falloppio (1523-1562) Italian physician	Described a tube which leads from the ovary to the uterus (Fallopian tube).
Andreas Vesalius (1514-1564) Dutch anatomist	**Father of Modern Human Anatomy.** Wrote "*On the Fabric of the Human Body.*" (1543),	Michelangelo Buonarroti (1475-1564) Painter, sculptor, architect, engineer and inventor	Greatest influence on Western art. Studied anatomy using corpses from a church hospital.

32 Jacob Brownowski in *Magic, Science and Civilization*, 1978.

Founding of Scientific Societies
Scientific societies were founded in order to communicate discovery, understanding and interests among scientists. The earliest societies were formed in Italy. The *Acedemia del Lincei* (1603) in Rome and the *Academia Secretorum Naturae* (1560) in Naples were the first in Europe. In 1622, the *Societas Ereunetica* was established in Germany and the Berlin Academy in 1700.

Late Renaissance Leads to the Age of Enlightenment (1520-1600)
The most prominent scientist of the Renaissance was **Galileo Galilei**, an Italian mathematician, astronomer, physicist and philosopher. His work earned him the title of the "**Father of Modern Science.**" See table 1.8. He *set standards for measurement so that his work was reproducible and he was willing to change his views if observations demanded*. Hallmarks of modern science are: standards of measurement, reproducibility of experiments and the ability of investigators to change their minds if the evidence dictates.

Galileo proved that bodies of greater mass will not fall to Earth faster than a body of lesser mass. He proved mass is independent of gravity. He made improvements to the telescope, perfected the microscope and improved the military compass. He **debunked**[33] the idea that the Moon was **translucent.**[34] He was the first to observe and describe Sunspots, the **phases**[35] of **Venus,**[36] **Saturn**[37] and its rings. He observed the **Milky Way**[38] to be a collection of stars in 1610, not just a faint milky, cloud-like band in the sky. He founded an entirely new science, *physics*. *Physics studies the relationship between matter and energy.*

Galileo's astronomical observations supported **Copernicus'**[39] **heliocentric**[40] view of the known universe. It was very controversial at the time and **diametrically**[41] opposed to Ptolemy's **geocentric**[42] theory supported by the Roman Catholic Church. In 1632, Galileo published <u>Dialogue on the Great World Systems, The Ptolemaic and Copernican</u>.

33   To expose as false.
34   Light passes through an object.
35   How light falls and varies in appearance on the surface of the planet or Moon.
36   The second planet from the Sun. It has been known since ancient times.
37   Although known since ancient times, Saturn's rings were first observed by Galileo in 1610.
38   The galaxy that the Earth is contained in. From the Greek–*galaxias*–milky.
39   Nicolaus Copernicus was a Polish astronomer and Catholic priest who provided proofs for the Earth revolving around the Sun.
40   Heliocentric theory is a model of the known universe with the Sun at its center.
41   Exactly the opposite.
42   Claudius Ptolemy (c. A.D.100-170) proposed the geocentric theory of the universe. His model of the universe placed the Earth at its center.

This got him into trouble. In 1633, he was tried by the **Inquisition**[43] for his views on the heliocentric theory of the universe, found guilty and placed under house arrest until his death.

Table 1.8. Scientists and artists of the Late Renaissance.

Scientist	Achievement(s)	Scientist	Achievement(s)
William Gilbert (1544-1603) English physician	The Earth is a giant magnet. Laws of magnetism.	Marcello Malpighi (1628-1694) Italian physician	Discovered capillaries.
Galileo Galilei (1564-1642) Italian mathematician	**"Father of Modern Science."**	Francesco Redi (1628-1697) Italian physician	Proved flies do not come from rotting meat.
Willebrod Snell (1580-1626) Dutch scientist	Discovered law of refraction, condensed as Snell's Law.	Robert Hooke (1635-1703) English scientist	Called rectangular boxes in cork "cells." Law of springs.
Blaise Pascal (1623-1662) French mathematician	(Pascal's Principle) Pressure applied to a fluid in a closed container is transmitted equally to every part of the fluid on the walls of the container.	Issac Newton (1643-1726) English physicist	Three Laws of motion. Built on Galileo's work.
Robert Boyle (1627-1691) English scientist	Boyle's Gas Law: If temperature is constant and the volume of a container is decreased, the pressure increases.	Antonie van Leeuwenhoek (1632-1723) Dutch naturalist	**"Father of Microbiology,"** first to publish observations of microbes.

The Enlightenment

The 1700s are referred to as "The Age of Enlightenment." *Reason was seen as the only authority.* Rational thinking was applied to all aspects of thought, including the study of nature. See table 1.9. The individuals that made up the cultural and scientific communities of Europe and the English colonies in North America supported adherence to reason as the primary authority for explanations of events and phenomena.

---

43   A system within the Roman Catholic Church that was used to prevent groups of people from breaking away from the Church. It was begun in France, but later expanded to lands under Church control.

Table 1.9. Scientists of the 1700s in Europe and some of their achievements.

Scientist	Achievement(s)	Scientist	Achievement(s)
Lazzaro Spallanzani (1729-1799) Italian priest and biologist	Proved standing broth does not make microbes. First proof against abiogenesis.	Edward Jenner (1749-1823) English surgeon	"Father of Vaccination."
Carl Scheele (1742-1786) Swedish chemist	Discovered oxygen about 1772.	Lavoisier (1743-1794) French chemist	"Father of Modern Chemistry." Stated the Law of Conservation of Mass in chemical reactions first.
Luigi Galvani (1737-1798) Italian physician	Muscles of dead frogs legs twitched if electricity was applied.	Jean-Baptiste Lamarck (1744-1829) French naturalist.	The theory of inheritance of acquired characteristics. Giraffes have long necks because its ancestors stretched it to get food on higher branches.
Alessandro Volta (1745-1827) Italian physicist	Invented first battery (Voltaic pile). Sent electric current over a distance.	John Dalton (1766-1844) English chemist and physicist	Developed modern atomic theory.
Carl von Linne (1707-1778) Swedish biologist and physician	He is considered the "Father of Modern Taxonomy."	Joseph Priestley (1733-1804) English scientist.	Discovered oxygen in 1774.
James Watt (1799) English	Prevails in court case and credited as inventor of steam engine.	Joseph Black (1728-1799) Scottish physician.	Discovered latent heat, specific heat and carbon dioxide.
Laura Bassi (1711-1778) Italian anatomist and physicist	First woman to officially teach at a university in Europe.	Caroline Herschel (1750-1848) German astronomer	Discovered three new nebulae in 1783. She discovered eight comets.

Technology and Invention in the 1800s

The 1800s saw rapid scientific and technological advances. Means of travel and communication were born and developed explosively: railways, steamships and automobiles. Below are some inventions of the 1800s that we would recognize today. See tables 1.10 and 1.11.

Table 1.10. A summary of inventions of the 1800s and early 1900s.

Year	Inventor and Nationality.
1813	Peter Durand invents the tin can. A tin can enabled foods to be kept over long periods of time and shipped over long distances. French.
1816	Rene Laennec invents the stethoscope. French.
1829	Louis Braille invents braille printing for the blind. French.
1834	Cyrus McCormick invents the first successful reaper for cutting and gathering crops. American.
1835	Samuel Morse builds the first American telegraph. American.
1853	Dr. Francis Rynd invents the hollow needle that became the hypodermic needle. Irish.
1855	Henry Bessemer invents process for mass producing steel. English.
1867	Alfred Nobel invents dynamite. Swedish.
1869	Thomas Edison invents the electric vote recorder. American.
1872	American. George Westinghouse invents air brakes. Greatly improved passenger and freight train safety. American.
1876	Disputed: Innocenzo Manzetti (It.), Antonio Meucci (It.), Alexander Graham Bell (Am.) and Thomas Edison (Am.) telephone. Alexander Graham Bell (Am) received the first American patent.
1878	Thomas Edison invents the phonograph and perfects electric light bulb. American.
1880	Thomas Edison patented a system for distribution of electricity. American.
1885	Gottlieb Daimler builds the world's first four-wheeled gasoline powered motor vehicle. German.
1886	George Westinghouse builds alternating current (A.C.) generator and distribution system. The same system is in use today. American.
1902	Willis Carrier invents the air conditioner. American.
1903	Wright brothers. First sustained gasoline powered flight. American.
1903	Willem Einthoven invents electrocardiogram. Dutch.
1908	Henry Ford devises the assembly line for the production of gasoline powered automobiles. American.

Table 1.11. Scientists of the 1800s and early 1900s in Europe, the United States, Canada and some of their major achievements.

Scientist	Achievement(s)	Scientist	Achievement(s)
Matthias Schleiden (1804-1881) German botanist	Concluded after many investigations that plants are composed of cells.	Martinus Beijerinck (1851-1931) Dutch microbiologist	First to use term "virus." **Father of Virology.** Discovered tobacco mosaic virus.
Charles Darwin (1809-1882) English naturalist	Published *On the Origin of Species by Means of Natural Selection.* Natural selection is the basic mechanism of evolution of new species.	Henri Becquerel (1852-1908) French physicist and Manya Skłodowska (Madame Curie) (1867-1934). Polish physicist	Discoverer of radioactivity along with Madame and Pierre Curie.
Theodor Schwann (1810-1882) German physiologist	"All animals are made of cells." Discovered neurolemmocytes (Schwann cells), discovered pepsin, coined the term metabolism.	Thomas Hunt Morgan (1866-1945) American geneticist	Genes are located on chromosomes.
Claude Bernard (1813-1878) French physiologist	Modern concept of homeostasis. Coined term *"milieu interieur."*	Karl Landsteiner (1868-1943) Austrian-American biologist and physician	Discovered A, B, O blood groups and Rh factor. Discovered the polio virus.
Ignaz Semmelweis (1818-1865) Hungarian physician	Hand washing by physicians will prevent puerperal fever (childbed fever, caused by staphylococci or streptococci).	Jules Bordet (1870-1961) Belgian physician	In 1906, isolated *Bordetella pertussis* in pure culture. Believed it was the possible cause of whooping cough.

Table 1.11. Scientists of the 1800s and early 1900s in Europe, the United States, Canada and some of their major achievements.

Scientist	Achievement(s)	Scientist	Achievement(s)
Florence Nightingale (1820-1910) English nurse, writer and statistician	**Founder of modern nursing.** Founded first nursing school. Used pie chart to illustrate sources of patient mortality in military and civilian life. She became known as "**The Lady With the Lamp.**"	George Washington Carver (1865?-1943) American scientist, educator, humanitarian and former slave	He developed hundreds of products from peanuts, sweet potatoes, pecans and soybeans.
Rudolph Virchow (1821-1902) German physician	*Omnis cellula e cellula.* ("Every cell originates from another an existing cell").	Max Planck (1858-1947) German physicist	Founder of quantum theory.
Louis Pasteur (1822-1895) French chemist	Final proof against spontaneous generation, Germ Theory of Disease, vaccination against rabies. Heating of beverages to prevent spoilage (Pasteurization).	Otto Loewi (1873-1961) German pharmacologist	Transmission of nerve impulses are chemical in nature.
August Weismann (1834-1914) German biologist	Inheritance of traits only passed on from parent to offspring through egg and sperm.	Robert Goddard (1882-1945) American physicist and inventor	**Father of Modern Rocketry.** Demonstrated a liquid fuel rocket could work in a vacuum of space.

Table 1.11. Scientists of the 1800s and early 1900s in Europe, the United States, Canada and some of their major achievements.

Scientist	Achievement(s)	Scientist	Achievement(s)
Gregor Mendel (1822-1884) Austrian monk and scientist	Discovered laws of heredity in pea plants. Published his findings in 1866.	William Rontgen (1845-1923) German physicist	Discovered X-rays (1895).
Joseph Lister (1827-1912) English surgeon	Developed antiseptic surgery.	Guglielmo Marconi (1874-1937) Italian inventor	Wireless telegraph.
August Kekule (1829-1896) German chemist	**Father of Organic Chemistry.** Explained role of the carbon atom in organic chemical reactions.	Niels Bohr (1885-1962) Danish physicist	Devised the planetary model of the atom. A small, positively charged nucleus surrounded by orbiting electrons. Called the **Bohr Model of the Atom.**
Robert Koch (1843-1910) German physician	Specific microbes cause specific diseases.	Frederick Banting (1891-1941) and Charles Best (1899-1978) Canadian physicians	First to extract insulin from pancreatic tissue in 1922.
Camillo Golgi (1843-1926) Italian physician	Identified the intracellular reticular apparatus in 1898, an organelle called a Golgi apparatus or Golgi body.	Albert Szent-Györgyi (1893-1986) Hungarian physiologist	Protein actin combines with myosin and ATP to cause muscle fiber contraction.
Ilya Mechnikov (1845-1916) Russian biologist	Discovered phagocytosis.	Raymond Dart (1893-1988) Australian anatomist	Unearthed first human ancestor in Africa in 1924.

Table 1.11. Scientists of the 1800s and early 1900s in Europe, the United States, Canada and some of their major achievements.

Scientist	Achievement(s)	Scientist	Achievement(s)
Ivan Pavlov (1849-1936) Russian physiologist	Classical conditioning.	Paul Domagk (1895-1964) German pathologist	First commercially available antibiotic (sulfonamidochrysoidine). Effective against streptococcus.
Emil von Behring (1854-1917) German physiologist	Discovered diphtheria antitoxin.	Alexander Fleming (1881-1955) Scottish biologist	Discovered penicillin in 1928.
Joseph Thomson (1856-1940) British physicist	Discovered the electron.	Ernest Rutherford (1871-1937) British chemist and physicist	Discovered and named the proton. Concept of half-life.
Orville Wright (1871-1948) Wilbur Wright (1867-1912) Americans	First sustained gasoline-powered flight (1903).	Hans Krebs (1900-1981) German physician	German physician. Identification of the citric acid or (TCA) cycle.
Peyton Rous (1879-1970) American physician	Some viruses can transmit certain types of cancer.	Vincent du Vigneaud (1901-1978) American biochemist	Discovered the structure of oxytocin and vasopressin.
George Whipple (1878-1976) American physician	Biermer's anemia (Addison's anemia) can be reversed if affected dogs are fed liver. Applied successfully to humans.	Enrico Fermi (1901-1954) Italian-American theoretical physicist	First nuclear reactor, major contributions to quantum theory, nuclear and particle physics and the first atomic bomb.
Albert Einstein (1879-1955) German-American theoretical physicist	Special and general theories of relativity, prediction of the deflection of light by gravity.	Luis W. Alvarez (1911-1988) American physicist	Proposed theory that a giant impact on Earth's surface by an asteroid caused the extinction of the dinosaurs.

Table 1.11. Scientists of the 1800s and early 1900s in Europe, the United States, Canada and some of their major achievements.

Scientist	Achievement(s)	Scientist	Achievement(s)
Oswald Avery (1877-1955) Maclyn McCarty (1911-2005) and Colin McLeod (1909-1972) American physicians and microbiologists	Published seminal paper in 1944 proving deoxyribonucleic acid (DNA) is genetic material in bacterial transformation.	Frederick Sanger (1918-) English biochemist	Determined amino acid sequence of insulin.
George W. Beadle (1903-1989) and Edward L. Tatum (1901-1975) American geneticists	Demonstrated how genes make enzymes ("one gene, one enzyme" hypothesis).	Howard Florey (1898-1968) and Ernst Chain (1906-1979) Australian pharmacologist and British chemist respectively	Developed penicillin into a practical medicine (1939).
Carl Anderson (1905-1991) American physicist	Discovered the anti-electron or positron. First proof of anti-matter.	Linus Pauling. (1901-1994) American chemist, biochemist, peace activist, author and educator	Determined the nature of the chemical bond: how atoms link up to form molecules in both living and non-living systems.
Henry Ford (1863-1947) American industrialist	Employed the first conveyor belt assembly line for the production of automobiles.	Charles Drew (1904-1950) American physician, surgeon and medical researcher	Blood banking. He discovered that plasma can be kept longer than whole blood. His model became the standard for the American Red Cross.

The Second World War Spurs on Scientific Achievement

The Second World War (1939-1945) was the most tragic period in modern times. This war produced over 60 million deaths and untold suffering, and at the same time produced a great number of advances in science and technology, but without a **commensurate**[44] advance in moral and humanitarian concerns by some nations. Some of these discoveries occurred prior to World War II and were developed further during the war. Others were invented in response to needs of the war, and some were developed near the end of the war in 1945. Penicillin, plastics, synthetic rubber, radar, television, electron microscopes, nuclear science, ballistic missiles, jet aircraft, sonar and computers are a few innovations of this period. See table 1.12.

Major scientific figures, Albert Einstein from Germany and Enrico Fermi from Italy, fled the Fascist countries and came to the U.S. before the war started. Both were crucial for making the atomic bomb used to end the war.

World War II was a major turning point for how the world's people viewed politics, culture and technology. This was a period that marked the end of colonialism and the rise of nationalism in many overseas possessions of European countries.

In the scientific community, we see teams of people working on a single problem. The electron cloud model of the atom is the result of collaboration of many scientists. Although Enrico Fermi and J. Robert Oppenheimer are considered "Fathers of the Atomic Bomb," thousands of people made contributions to the project.

---

44   Having an equal amount.

Table 1.12. Major scientific figures and their achievements of the mid-1900s to the present in Europe, the United States and Canada.

Scientist	Major Achievement(s)	Scientist	Major Achievement(s)
Rachel Carson (1907-1964) American marine biologist	*Silent Spring*, published in 1962, made Americans aware of the dangers of using pesticides.	Carl Woese (1928-2012) American microbiologist	Proved archaea are genetically different from bacteria. Devised a three-domain system of classification.
Harold C. Urey (1893-1981) and Stanley L. Miller (1930-) American scientists	Duplicated conditions of a primitive Earth *in vitro* that produced simple organic compounds (1953).	Eric Kandel (1929-) American neuroscientist	Memory is stored in neurons.
Robert Whittaker (1920-1980) American plant ecologist	Devised a five kingdom system of classification for living things.	Paul Berg (1926-) American biochemist	Pioneer in genetic engineering.
Robert Edwards (1925-2013) British biologist	Developed *in vitro* fertilization.	Gerald Edelman (1929-) American biologist	Discovered antibody structure.
Francois Jacob (1920-2013) along with Jacques Monod (1910-1976) French biologists	Demonstrated mechanism of gene repression.	Luc Montagnier (1937-) French virologist	Discovered the Human immunodeficiency virus (HIV).
Joshua Lederberg (1925-2008) American molecular biologist	Discovered that bacteria can mate and exchange genes.	Stanley Prusiner (1942-) American neurologist	Discovered prions (infective proteins).
Paul Ehrlich (1854-1915) German physician	**Father of Modern Chemotherapy.**	Jonas Salk (1914-1995) American medical researcher and virologist	Developed first polio vaccine.

Table 1.12. Major scientific figures and their achievements of the mid-1900s to the present in Europe, the United States and Canada.

James Watson American (1928-) and Francis Crick English (1918-2004)	Discovered the structure of the DNA molecule in 1953.	Rosalind Franklin (1929-1958) British biophysicist	Her X-ray diffraction images helped to lead to the discovery of the structure of DNA.
Dmitri Mendeleev (1894-1907) Russian chemist	Devised the Periodic Table of the Elements.	Murray Gell-Mann (1929 -) American physicist	In 1960s, he determined that protons and neutrons are made up of smaller particles, quarks.
Clair Patterson (1922-1995) American geologist	First to determine in 1953 that the Earth is 4.6 billion years old.	John Glenn (1921-) American astronaut	First to orbit the Earth (1962).
Yuri Gagarin (1934-1968) Russian astronaut	Soviets launch first man in space. (1961).	Buzz Aldrin (1930-) and Neil Armstrong (1930-2012) American astronauts	Apollo astronauts land on the Moon (1969).
Tim White, American paleontologist (1950-) Berhane Asfaw, Ethiopian paleontologist and Gen Suwa, Japanese paleontologist	In 1993, the team found parts of a hominid skull, jaw, and arm bones in Ethiopia, Africa. The bones dated back to 4.4 million years ago.	Centers for Disease Control (CDC)	AIDS officially recognized to be caused by a virus in 1981.
Edward Roberts (1941 - 2010) American engineer	Introduced personal computer as a kit, the Altair 8800 in 1975.		

Chapter 2 Science and the Scientific Method
The Natural Sciences

Natural science (science) is the study of the entire natural world. The natural world consists of all living and non-living things on Earth, the solar system and the universe. Clearly, all of nature is too broad an area for any one person to study. Science was subdivided into two general areas of study over time: biological science and physical science. **Biological sciences** explore all aspects of living things. It includes the study of **animals,**[45] **plants**[46] and **microscopic living things.**[47] *Physical sciences* investigate the non-living world of chemical reactions, how energy and matter are related, the rocks, water, and atmosphere of the Earth and **celestial**[48] bodies. Even these subdivisions are too broad for one person to study. Each is subdivided further into smaller areas in order to be studied efficiently. See figure 2.1.

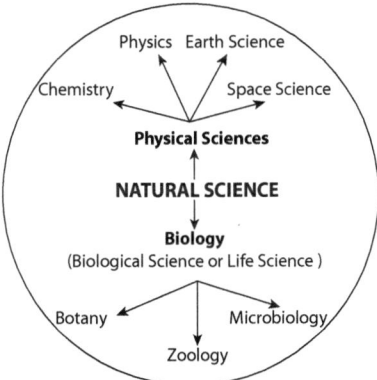

Figure 2.1. Diagrammatic representation of the major branches of natural science.

Individual physical and biological sciences influence and overlap each other. There are no hard and fast boundaries. As an example, the science of biochemistry requires knowledge of biology and chemistry. Biophysics is the use of physical laws to understand biological systems.

Space and Earth Sciences

Space science is the study of everything beyond the Earth's atmosphere. It examines how the universe began, how the Earth formed and how the stars and planets were formed. Biology and space science intersect when we explore the possibility of life on other planets.

The universe formed about 12 to 14 billion years ago. That was the time of the "big bang," the moment the universe was created. The **solar system**[49] condensed from a giant cloud of interstellar dust about 4.5 billion years ago. It consists of the planets and other materials in space that are under the influence of the gravitational pull of the Sun.

45   Animals are multicellular, eukaryotic organisms.
46   Plants are multicellular, eukaryotic, chlorophyll-bearing organisms that carry on photosynthesis.
47   Primarily one celled archaea, bacteria and protozoa, but can be microscopic multicellular animals such as dust mites.
48   Concerned with stars, planets and other bodies in the universe.
49   The Sun and the bodies that rotate around it.

Earth science is the study of non-living things on the Earth: its **atmosphere**[50] (see table 2.1), **hydrosphere**[51] and **lithosphere**.[52] Geology is the study of the Earth's rocks and the processes that enabled the Earth to form and change over time. The atmosphere is 79% gaseous nitrogen (N_2), 20% gaseous oxygen (O_2) and smaller percentages of other gases.

Table 2.1. Composition of the Earth's atmosphere.

Gases or Substance	Symbol or Formula	Percentage
Nitrogen	N_2	78.08
Oxygen	O_2	20.95
Water vapor	H_2O	0 to 4
Carbon dioxide	CO_2	0.038
Argon	Ar	0.93
Neon	Ne	–
Helium	He	–
Krypton	Kr	–
Nitrogen oxides	–	–
Sulfur compounds	–	–
Ozone		

Humans live on the solid Earth bathed in a gaseous atmosphere. They use molecular oxygen (O_2) for cellular respiration. The major function of O_2 is to accept harmful products of metabolism.

The Biological Sciences

Biology is the study of living things. Living things possess special properties or life characteristics. Only living things exhibit these characteristics. If something displays life characteristics, then it is a life form, and it is studied by a biologist.

Life on Earth began about 3.4 billion years ago as simple prokaryotic cells. The genus *Homo* first appears in the fossil record about 2 million years ago. *Homo sapiens*, modern man, appears in the fossil record about 100,000 to 200,000 years ago.

Biology consists of three major branches: *zoology, botany* and *microbiology*. See figure 2.1. Each of these three areas of study is general in their own right, but more specific than biology. Each, in turn, is broken down into smaller, more specialized areas of study.

---

50   The air living things breath.
51   The Earth's waters.
52   The Earth's crust.

Zoology is the study animals. Animals are eukaryotic and multicellular. Eukaryotic means their cells have a true nucleus. A true nucleus is one that is surrounded by a double-layered membrane. Their mode of nutrition is heterotrophic. Animals obtain their food from the environment. **Anatomy,**[53] **physiology,**[54] **histology,**[55] **ornithology,**[56] **ichthyology,**[57] **herpetology,**[58] **entomology,**[59] embryology and comparative zoology are a few branches of zoology.

Botany is the study of plants. Plants are eukaryotic, multicellular, chlorophyll containing organisms that carry on photosynthesis. Their mode of nutrition is *autotrophic*. Autotrophs make their food from carbon dioxide and water. **Phytotomy,**[60] **plant physiology, phytopathology,**[61] **agronomy,**[62] **plant ecology** and **plant genetics** are a few divisions of botany.

Microbiology is the study of life forms that cannot be seen without a microscope. Subdivisions of microbiology are: bacteriology, virology, **algology,**[63] **mycology,**[64] protozoology, and parasitology. *These six subjects define the science of microbiology,* even though some organisms studied may be **macroscopic,**[65] such as mushrooms, many parasitic worms and some algae. Most microbes are prokaryotes. Some are eukaryotes. Prokaryotes do not have true nuclei because their DNA is *not* surrounded by a double-layered membrane.

Viruses and *prions* are often studied in a general microbiology course, even though they are not living entities. All viruses are parasites. They consist of a nucleic acid core, DNA or RNA, surrounded by a protein coat. Viruses need to live and reproduce in a living cell. Prions are improperly folded proteins that infect and cause pathology in humans. Stanley Prusiner coined the term prion in 1982 to describe the agent of bovine spongiform encephalopathy ("mad cow disease"). Unlike other infectious agents, prions do not contain nucleic acids.

---

53   The study of structure of living things.
54   The science that studies the function of living things.
55   The science of tissues.
56   The study of birds.
57   The study of fish.
58   The study of amphibians (frogs and toads) and reptiles (snakes, turtles, lizards and crocodiles).
59   The study of insects.
60   Plant anatomy.
61   Diseases of plants.
62   The study of crop plants such as corn, wheat and other grains.
63   The study of algae.
64   The study of fungi.
65   Objects that is visible to the naked eye. From the Greek– macro meaning large.

The Physical Sciences
Physics

The word "physics" is derived from the ancient Greek word *physis* meaning "nature." Modern physics is a branch of natural science that deals with the non-living world. Physics studies the interrelationship between **matter**[66] and **energy**.[67] Originally termed *natural philosophy*, the physics of the ancient Greeks was an exploration into the workings of nature. Thales' (624-c. 546 B.C.) was the first philosopher that tried to explain natural phenomena without invoking the intercession of gods or some other supernatural force. He believed that there was a natural cause for events. Thales' ideas resurfaced during the scientific revolution that took place during the Italian Renaissance from the late 1200s to 1500.

The science of physics studies how matter and energy are interrelated. One of the most famous examples of this relationship is Albert Einstein's famous equation $E = mc^2$. In the preceding equation **E** represents energy, **M** represents **mass**[68] and **c** equals the speed (v) of **light**.[69] The validity of this equation was proven when a piece of uranium about the size of an apple was converted into massive amounts of energy at the Trinity Test Site at the White Sands Proving Grounds in New Mexico in 1945.

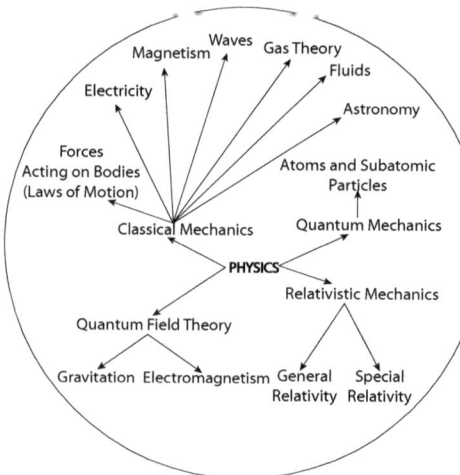

The science of *Physics* has made outstanding beneficial contributions to modern technology. Physics is also divided into many sub-sciences. **Classical mechanics**,[70] gravity, light, lenses, sound, electricity, magnetism and more recently **quantum mechanics**[71] are the most common divisions of the science of physics. *Biophysics* is an overlapping study between biology and physics.

Figure 2.2. Major divisions of physics.

---

66 There are four fundamental states of matter: solid, liquid, gas and plasma.
67 Energy is defined as something that has the capacity to do work.
68 Mass is defined as the amount of matter in an object.
69 In a vacuum it is 186,000 miles per second (299,792,458 meters per second).
70 The study of how matter (bodies) react when a force is applied to them.
71 Studies the nature of particles, waves and the interaction of matter and energy.

The Scientific Method

The concept of science began with the ancient Greeks. They laid the foundation for modern science that grew and bloomed in Italy between the later Middle Ages (late 1100-1200s) and the height of the Italian Renaissance (c.1480-1520).

Greek science was based on discussion and logic, not experiment. Some Greek thinkers believed in *natural causes for events;* a very modern thought. They believed logic could lead to the discovery of truths about nature, if the **premise**[72] is correct, but if the premise is wrong, then what follows will be wrong. Democritus (460-370 B.C.) reasoned that *if* a sample of a substance such as silver was repeatedly cut of in half, *then* eventually the substance could not be cut in half again and still be the same substance. The piece is uncuttable. The ancient Greek word for uncuttable is *átomos*. His premise was correct. Democritus arrived at a model of the atom 2,500 years ago through reason.

Galileo believed *truth can only be obtained through* **experiment**[73] *and each experiment must be* **verifiable**[74] *by other scientists.* Modern scientists would recognize Galileo's method because it is his method we use today. There are variations of the scientific method, but all can be reduced to what I call five easy pieces:

1. **Question** – A statement about an observed event designed to reveal its cause.

2. **Hypothesis** – A tentative solution or answer to a question.

3. **Experiment** – A test of the truth or falsity of a hypothesis. An experiment consists of two groups:

   a. The *control group* contains all the variables except the *independent variable*, the variable under investigation.

   b. The *experimental group* contains the variable under investigation, the independent variable, as well as all variables of the control group. *Constants* are variables common to the control group and experimental group.

4. **Results** – A result is a measurable effect of the independent variable. It is called the **dependent variable**.[75]

5. **Conclusion** – Conclusions are made by analysis of the results and reasoning to a conclusion about the original hypothesis. Is it correct or incorrect? Which ever it is, it must be explained in detail.

---

72 A statement about something that is assumed to be true.
73 The word experiment comes from the Latin–*experīmentum* meaning "test," "proof" or "trial." In French it is identical to the word experience.
74 Provable or testable.
75 The independent variable is the variable controlled by the experimenter. It is what is being manipulated.

Details of of the Scientific Method
1. Question
The scientific method begins with a *properly* formulated question. If the wrong question is asked, the answer will be useless. *The question must be constructed in such a way that one variable is isolated.* Columbus asked, "Can the east (China) be reached by sailing west?" He had investigated the problem until he believed he could solve it. He is given credit as the discoverer of the "New World" because he published his work and like other scientists and explorers he established **priority**.

2. Hypothesis
A *hypothesis*, a *tentative solution to a problem,* is formed after becoming familiar with information about a problem. Columbus believed that the Earth is round. He hypothesized that sailing west will bring him to China or Japan in the East and thence back his starting point. Some people thought the same in times past, but he was the only one to carry out an experiment based on his firm belief in his hypothesis. He sailed west to reach the east and in doing so discovered the New World. He just did not know about the two continents between Europe and Asia.

3. Experiment
An ***experiment*** is a method of gathering information about an observable event in nature. An experiment is a *test of a hypothesis*. The experimental design must *force* a situation to occur that will duplicate conditions as they occur in nature. The design must control for, and *isolate, only one variable,* and eliminate all others.

Aristotle believed that heavy objects fall faster than light objects when dropped from a height. His belief was based on argument *not* experiment. His idea became **dogma**[76] for 2,000 years because of his authority. Galileo challenged Aristotle's conclusions and authority. Galileo hypothesized that all objects, when dropped from a height at the same time, regardless of mass, will hit the ground at the same time. He asked the question: "Do objects of different masses fall to Earth at the same *speed* or different speeds when dropped from the same height? There were no stop watches. He used his pulse as a timing device and made a very long inclined plane (about 30 feet) with a smooth channel running down its length. He allowed perfectly round brass balls of different masses to roll down the inclined plane and timed how long it took each ball to reach the bottom of the ramp. All of the different masses arrived at the bottom of the ramp at the same time.

---

76   An absolute truth.

4. Results

With careful measurements, Galileo's results demonstrated that balls of different masses arrived at the bottom of the inclined plane at the same time because they had the same rate of **acceleration**.[77] Equally important, his work was reproducible. He carefully recorded and analyzed the data. Thus, a 2,000 year old belief was disproved by experiment.

5. Conclusion

Based on his careful measurements, Galileo concluded that the mass of a body was independent of gravity as it falls to Earth. It made no difference if the mass was big or small. Therefore, gravity did not affect the speed of a falling mass. Galileo published his work for other scientists to review and repeat. After analyzing the **data**,[78] the scientist must determine if the hypothesis it correct or incorrect. It will not simply be a "yes" or "no" answer. This is the conclusion and must be supported by the scientist's own **criticism**[79] of the experiment. A thorough investigator tries to destroy the hypothesis to see if it continues to hold true.

Science is the study of nature. Scientists ask questions about observed **phenomena**.[80] For example: "What causes pneumonia?" "How is rain formed?" "What is the cause of global warming?" Scientists want to understand the *cause* of an *effect*. For example, an observation is made that plants grow taller in one area of a field as opposed to another (an effect), even though all other conditions such as rainfall, light and type of plant are the same. Analysis of the soil reveals a higher content of sodium nitrate ($NaNO_3$). The question is asked, *"Does $NaNO_3$ cause better growth of these plants?"* A *hypothesis* is proposed: "I believe sodium nitrate ($NaNO_3$) increases the rate of growth of grass plants." A simple *experiment* is designed to prove or disprove the hypothesis. The compound $NaNO_3$ (*independent variable*) is added to only one of two trays of identical grass plants. Soil, water, light and oxygen are the same in both trays. These are the *constants*. If the $NaNO_3$ is responsible, then the effects of $NaNO_3$ should be clearly visible in the height (*dependent variable*) of plants growing in the tray containing $NaNO_3$. If the results are an increased rate of growth in the $NaNO_3$ containing tray, then the *conclusion* would be stated that indicates the validity of the hypothesis. The application of the scientific method to answer questions may produce new or strengthen old theories, laws or facts.

77   Coined by Galileo, "acceleration" from the Italian–meaning *additional speed*.
78   Data (*pl.*) is either a measured quantity or a quality such as a feature of something like color. Typically data are quantities expressed as measurements
79   Criticism is a good thing. It deals with commenting on the "good" and the "bad" aspects of a particular action or creation. This is how things are improved.
80   Any observed event.

Forms of Trustworthy Scientific Information
Theories, Laws and Facts

Students of science should know the distinction between *theories*, *laws* and *facts*. These are three kinds of trustworthy scientific information used to express scientific knowledge.

Theories explain phenomena in nature. Theories are broad generalizations that are accepted as true until they are proven false. The theory of **evolution**,[81] for example, has new discoveries added on occasion. As a result, the theory is strengthened or modified. For example, Walter Alvares hypothesized that the impact of an asteroid on the Earth's surface wiped out the dinosaurs. His hypothesis is known as ***catastrophism*** (***impact hypothesis***). Initially rejected, its **adherents**[82] verified new facts that supported this hypothesis. The alternative view of catastrophism is the hypothesis of ***uniformitarianism***. Uniformitarianism states change was very gradual because the laws that govern nature never changed. New evidence did not support this hypothesis. Uniformitarianism held sway for about 150 years, but was discarded and replaced by catastrophism, a hypothesis that was proven to be true. This is one hypothesis in the overall theory of evolution.

Laws define or describe a phenomenon with mathematical precision. For example, Newton's Laws of Motion describe how a mass will act under different conditions. If a ball is thrown into the air near the surface of the Earth, its position can be predicted anywhere along its path, how long it will take to fall back to Earth and exactly where it will land. The scientific community will accept this verifiable statement as a law. Although laws are accepted as true, they are still reexamined, retested and may need to be extended, modified or discarded.

A *fact is the result of an experiment*. Hypotheses are proposed and confirmed by experiment. The fact then enters into the body of scientific knowledge. Experimental findings are always challenged. If a finding is proven wrong, then other hypotheses are considered and tested experimentally.

Forms of Information That Are Not Trustworthy
Anecdote

Anecdotal information is a result of casual, nonscientific observations passed from person to person. It could be information that is true or not true. It could be trustworthy or not. *It is not peer-reviewed information published by members of the scientific community.* It is not statistical proof. It is hearsay.

81 Changes in a population over time. Changes that accumulate over generations.
82 Those that stick or cling to an idea.

Science is a search for truth. Sadly, some individuals engage in behaviors that are in opposition to the scientific community's ideals. Below are some serious behaviors that risk the advancement of science and endanger the body of scientific information.

Scientific Integrity[83]
A breach of scientific integrity may fall into one of two general categories: *research misbehaviors* and *research misconduct*.

Research Misbehavior
Research *misbehavior* is a serious offense. If a researcher plagiarizes material or falsifies data, the researcher is guilty of misconduct. The principle investigator (PI) is guilty of misbehavior because he did not properly supervise research.

Research Misconduct
A *standard code* exists to guide scholars in their pursuit of knowledge. Violation of this code is termed *scientific misconduct*.[84] Misconduct can take three major forms: plagiarism, falsification and fabrication. Such conduct is a blatant disregard of for the ethical standards of the scientific community.

Plagiarism
Plagiarism is one of the most common forms misconduct. *Plagiarism is taking credit for written work that belongs to someone else.* It is stealing. An author may "forget" to cite another person's work. This should not happen if the investigator is *honest* and *thorough*. This is a purposeful act in the great majority of cases.

Falsification
Adding something that is not ordinarily present or removing data that should be present is *falsification*.

Fabrication
Piltdown Man is one of the most widely known scientific *fabrications*[85] of data in the history of **anthropology**.[86] The fabricator, a well-respected scientist, combined a jawbone of an orangutan with the skull of a modern human to approximate a "missing link" between apes and man. He then planted the "find" in a gravel pit in Piltdown, England in 1912. He then led a group of respected anthropologists to the site and allowed them to "discover" the greatest find of all time. Modern scientific techniques exposed the hoax in 1953.

---

83   Keeping moral and ethical principles in any endeavor, soundness, honesty and having moral character.
84   As defined by the National Science Foundation.
85   Make something with the intent to deceive.
86   Anthropology is the study of humanity.

Making False Claims by Pretending to be Scientific
Pseudoscience

Pseudoscience or false science is characterized by vague claims and lacks scientific verification. Pseudoscience is characterized with making wild claims for a remedy and linking the fraudulent product to a famous person.

The medical field is rife with examples of pseudoscience because there is a lot of money to be made by selling miracle drugs. One of these concoctions from a book of homeopathic medicines from the 1880s went as follows: powdered starfish, secretions from a skunk, powdered coal, human urine, which supposedly cured all illness, a **panacea**.[87] These concoctions were taken in small amounts and were generally harmless.

---

87   Named after the Greek goddess of universal remedy.

Chapter 3 Measurement
Measurement Defined

Measurement is matching a number to an unknown physical amount of something. This magnitude (number) or "how much" is represented by a standard. A *standard* is a known amount of something that is used as a basis of comparison against an unknown amount of something. For example, if a standard 1 kilogram mass is placed on one side of a double pan balance and salt is poured on the opposite side until the kilogram mass is balanced by the pile of salt. We have **massed**[88] an exact amount of salt equal to 1 kilogram. *One kilogram is the magnitude. Salt is the physical "something."*

Standards for length, mass and volume existed in the ancient world. Peoples of ancient China, Egypt, Mesopotamia, and the **Indus**[89] valley used very accurate standards of measurement for their commerce. The Egyptian cubit, a unit of length, was the distance from the elbow to the tip of the furthest digit. A standard cubit was made of stone and copies were made of wood for portability.

Standardization on a massive scale is first found in lands administered by the Romans (700 B.C.-500 A.D.). Rome traded with many provinces beyond Italy. Standards were agreed upon and universally adopted in the Roman world. The Roman mile, equal to 1,000 paces of a Roman soldier, was the standard mile for civilian and military transportation. All distances were measured according to this standard. The English word "mile" is a contraction of *mille passus* (L. 1,000 paces). The *ligula* was a Roman unit of volume equal to about 11 ml and the *libra*[90] (lb.) about 329 g. People went back to their native systems of measurement after Roman administration ended. Commerce became difficult.

The period of the French Revolution (1879-1899) saw the creation of a new method of weights and measures. The French wanted new standards of measurement based on something natural. The base unit for length, the *meter*, see figure 3.3, was defined as one ten millionth (1/10,000,000) of the Earth's circumference from the North Pole to the equator. The meter was used to construct the liter for volume and the gram for mass. This was an easy to use, simplified decimal system in which base units are multiplied or divided by 10. It was formally adopted in 1795 and named the *metric system*. Copies of the standard meter, liter and kilogram were distributed throughout France.

---

88 "Massed" because an unknown amount of matter is being compared to a known amount of matter, such as the kilogram. Mass is the amount of matter in an object.
89 A region bordering the Indus river in India (2500-1800 B.C.).
90 A little over seven-tenths of a pound.

By 1875, the modern metric system, the **International System of Units**[91] or the **SI system**,[92] was *agreed upon* as a system of measurement by all industrialized countries, except Britain. Rulers, balances and graduated cylinders are calibrated to the SI system. They are used in laboratories around the world and in marketplaces. By 1900, the metric system was in use throughout Europe and Latin America, fulfilling the need for *uniformity* of measurement.

Measuring devices have a *scale* imprinted on them. *A scale is a line divided into equal parts.* Each division represents one unit of the base unit. A scale can be crudely made or very fine. If the scale is fine, then precision of the device will be increased. If measurements are done carefully, they become more accurate. There are seven base units in the modern metric system. Each standard is represented by the quantity 1. Standards are subdivided or multiplied into units for ease of use.

A *unit is a defined or definite amount of a physical quantity.* The standard meter, liter and kilogram, and five other standards of the SI system are kept in a governmental department that is responsible for weights and measures in every country. In the United States, the standards are kept in the National Institute of Standards and Technology of the United States Department of Commerce.

The *seven base units of the SI system* are the agreed upon standards against which everything is measured. The base units in the SI system are the: **meter**[93] (m), **kilogram (kg)**,[94] **second (s)**, **ampere (A)**, **kelvin (K)**, **candela (cd)** and the **mole (mol)**. Multiples and fractions of base units are preceded by a **prefix**.[95]

The most commonly used base units are the *meter* (m) for length and the *kilogram* (kg) for mass. The kilogram replaced the gram of the early metric system as the standard for mass. The cubic meter (m^3), a unit of volume, is not a base unit and is too large for everyday use. A convenient unit of volume to use is 1,000 cm^3 **(1,000 cc)**.[96]

To measure something smaller than a meter, we have to use fractional parts of the meter such as centimeters and millimeters. To measure things larger than a meter, multiples of the meter are used such as a decameter or kilometer. Base units are preceded by prefixes that indicate fractions of or multiples of 1 standard unit. Commonly used prefixes are listed in table 3.1.

---

91 *Système International d'Unités*, abbreviated SI.
92 All countries use the SI except the U. S., Liberia and Myanmar (Burma).
93 The meter is equal to 39.37008 inches.
94 The base unit of mass originally was a gram. It was replaced by the kilogram.
95 A word can be modified when a prefix is placed in front of it.
96 Cubic centimeter may be abbreviated cc. Commonly used in medicine.

Table 3.1 Most frequently used prefixes in the SI system of measurement.

Prefix	Number	Fraction	Decimal Equivalent	Scientific Notation
fempto	100,000,000,000,000	1/100,000,000,000,000	0.000000000000001	1×10^{-15}
pico	100,000,000,000	1/100,000,000,000	0.000000000001	1×10^{-12}
nano	1,000,000,000	1/1,000,000,000	0.000000001	1×10^{-9}
micro	1,000,000	1/1,000,000	0.000001	1×10^{-6}
milli	1,000	1/1,000	0.001	1×10^{-3}
centi	100	1/100	0.01	1×10^{-2}
deci	10	1/10	0.1	1×10^{-1}
Standard Unit 1 Meter, 1 Liter and 1 Kilogram ($1 = 1 \times 10^{0}$)				
deka	10	na	na	1×10^{1}
hecto	100	na	na	1×10^{2}
kilo	1,000	na	na	1×10^{3}
mega	1,000,000	na	na	1×10^{6}
giga	1,000,000,000	na	na	1×10^{9}
tera	1,000,000,000,000	na	na	1×10^{12}

Length
Length is defined as the distance from one location (A) to another location (B). It is used to measure with the *standard meter* (**m**). The meter is divided into 100 cm and each cm is divided into 10 mm. Therefore, 1,000 mm are in 1 m.
The meter was originally agreed upon as 1/10,000,000 of the distance from the North Pole to the Earth's equator. With advances in the science of measurement or *metrology*, the definition of the meter is now the distance light will travel in a vacuum in 1/299,792,458 seconds.

Volume
The amount of space an object takes up is its volume. If an object is **regular**[97] in shape, like a cube or a rectangular box, then the volume can be calculated by using the formula: $V = L \times W \times H$
V = volume, L = length, W = width and H = height

If L, W and H are measured in **meters**, V will be expressed in m^3
If L, W and H are measured in **decimeters**, V will be expressed in dm^3
If L, W and H are measured in **centimeters**, V will be expressed in cm^3

For example: $10 \, cm^1 \times 10 \, cm^1 = 100 \, cm^2$
$100 \, cm^2 \times 10 \, {}^1cm = 1,000 \, cm^3$
When multiplying exponents, we add them. **Do not write** cm^1, because cm to the power of 1 is written as cm. It is being done here for instructional purposes only.

The cubic meter is the unit of volume in the SI system. It is *derived* from the meter. The cubic meter measures 1 m x 1 m x 1 m. It is not a practical unit to use in the laboratory or the marketplace because it too big. The smaller cubic decimeter (dm^3) is more convenient to use.

Deriving the ml or cm^3 from the meter
If a cube measures 1 dm x 1 dm x 1 dm then the cube has a volume of 1 dm^3, which is equal to 1,000 cm^3 because a dm contains 10 cm. If 1/1,000th of 1,000 cm^3 equals 1 cm^3 (cc) and it was decided to call 1,000 cm^3 1 liter (L) then 1/1,000th of a L equals 1 milliliter (mL).

Simply put: **1 dm^3 = 1,000 cm^3 = 1 L**
Therefore: **1 cm^3 = 1 mL**

---

97   An object that has symmetry is a regular object. It has straight edges.

The volume of a liquid is measured with a graduated cylinder. If a box is constructed that measures 10 cm x 10 cm x 10 cm and filled with pure water at 4 °C, it has a volume of 1,000 cm³. The 1,000 cm³ of water is poured from the cube into a glass cylinder. If a mark is placed on the glass cylinder at the highest level of the water, the volume of the water is equal to 1,000 cm³ or *1 liter*. If the cylinder is divided into 1,000 equal divisions, then each division is equal to 1 milliliter (mL). See figure 3.1.

A slight downward bend of the upper surface of a column of water can be seen in a graduated cylinder. This is the *meniscus*. *The number of* mL *should be read at the bottom of the meniscus*. This distortion is compensated for by a bend in the bottom of the glass cylinder. See figure 3.1.

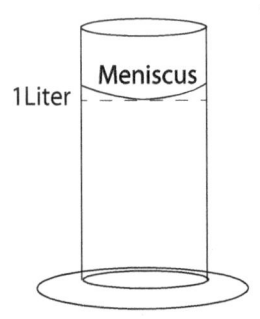

Figure 3.1. A graduated cylinder

Mass

Mass is the amount of **matter**[98] *contained in an object.* If the matter is a piece of iron, then the mass of the iron is the amount of iron atoms present in the sample. If the mass is a kilogram of water, then the total number of water molecules make up its mass. When balances are used, objects are being *massing* not weighed. This is because the mass of the unknown object is being compared to the standard mass called a *kilogram* we ave all agreed upon to use. Balances are calibrated to this standard mass. One thousand cm³ (1 liter) of pure water at **1 atmosphere**[99] and a temperature of **4° C** has a mass (m) of 1 kilogram.

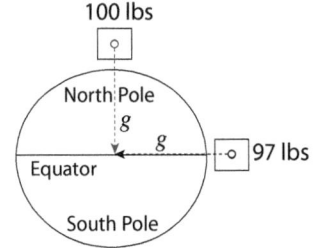

Do not confuse mass and weight. Mass does not depend on the pull of gravity. The amount of matter in an object does *not* change from location to location.

Figure 3.2. The effect of gravity at two different positions on Earth.

Weight does depend on the pull of gravity. *Weight is a measure of the pull of gravity on an object.* Since gravity changes slightly with position on the **Earth**,[100] so will weight. See figure 3.2.

---

98   Matter may be an element or a compound and occurs commonly in three different states: solid, liquid, gas.
99   1 atmosphere is the pressure exerted by the atmosphere on objects at sea level.
100   Since the Earth is flattened slightly at the poles and bulging slightly at the equator, a person at the equator will be further from the center of the Earth and will weight slightly less than at the poles.

Estimation

Measurement in the laboratory commonly involves using rulers, graduated cylinders and balances. None of these devices is perfect. That is why we have to approximate measurements frequently. An approximation of a measurement is called **estimation**.[101] The measuring unit or division marks on an instrument are matched or calibrated to a standard, such as a standard meter, kilogram, or liter. We have to estimate a number when measurement lies between two lines on a scale, but the estimation cannot exceed the limits of precision of the instrument being used. See figure 3.3. We can only be accurate to a centimeter because centimeters are the only divisions on the scale on this ruler.

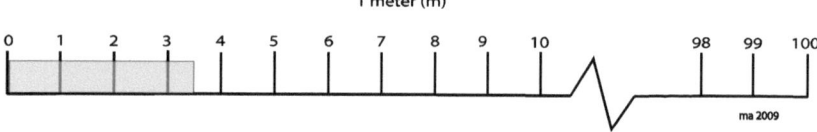

Figure 3.3. A standard meter consists a of 100 centimeters.

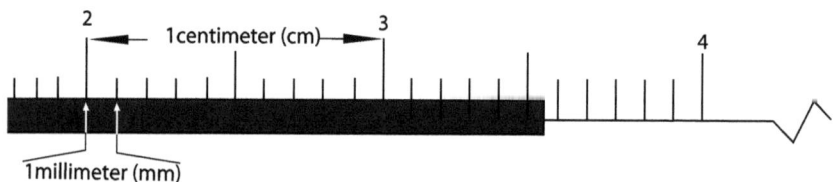

Figure 3.4. A portion of a meter ruler showing two centimeters. Each centimeter is divided into ten millimeters.

Precision and Accuracy

It is necessary that we use precise instruments to measure physical quantities. **Precision** is a function of the measuring device. **Accuracy**[102] is a function of the individual doing the measuring and how close the measurement is to the accepted value. Accuracy is how close an individual comes to the accepted value of what is being measured. Precise instruments are good, but if the person is careless, then inaccurate measurements will result. If the instrument is crude, then no amount of care will produce an accurate measurement. Clearly, the metric ruler in figure 3.4 is more precise that then that in figure 3.3.

Careless measurements will not be accurate. Accuracy can be improved with practice. No amount of practice will increase accuracy with imprecise instruments.

---

101   We sometimes have to estimate a measurement between two lines on a scale.
102   Accuracy refers to how close a measurement is to the actual value.

Additional Units of the SI System of Measurement

Time is the measurement of how long a particular event lasts or the interval between events such as the interval between waves striking the shore. A thunderstorm may last one hour or it may take three days for a seed to sprout. The *second* (**s**) is the basic unit of time in the SI system. Multiples and divisions of the second are arrived at by using the prefixes in table 3.1 so that a kilosecond, for example, would be 1,000 seconds and a microsecond would be 1,000,000th of a second. We commonly use minutes, hours, days, weeks and years in science, although these units of time are not part of the metric system.

Temperature is a measurement of the average kinetic energy of the atoms or molecules in an object. All atoms and molecules, whether they be solids, liquids or gases are in constant motion due to their kinetic energy. The faster atoms or molecules move, the further apart they will be spaced. *The SI unit of temperature is the kelvin (K).* In most laboratories and everyday use around the world the degree centigrade is used.

Water molecules in the *solid state* (ice) are *moving slowly*. In the *liquid state*, water molecules *move faster* than in the ice state and in the *gaseous state* water molecules *move fastest*. As the temperature in a substance rises, its atoms or molecules move further apart. Conversely, as the temperature goes lower, the atoms or molecules move closer together. When a nurse takes the temperature of a patient, the thermometer is measuring the average kinetic energy of the liquids it is in contact.

Heat is the transfer of thermal energy from one system to another. ***Thermal energy*** is the *total energy present in a body* caused by the random motion or kinetic energy of the atoms or molecules in the body.

If a 1 kg mass of iron at a temperature of 50 °C is placed on a block of ice (0 °C), heat will be transferred from the 1 kg mass of iron to the 1 kg block of ice and melting will occur. The thermal energy in a 1 kg cube of iron is the total kinetic energy all of the iron atoms in the 1 kg of iron. The thermal energy in a kg of ice is the total of the kinetic energy of all the water molecules in the 1 kilogram block of ice. See figure 3.5

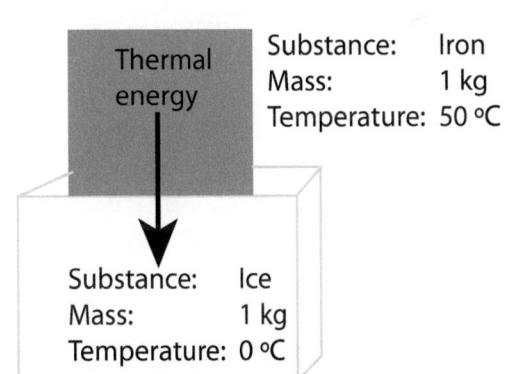

Figure 3.5. Transfer of thermal energy.

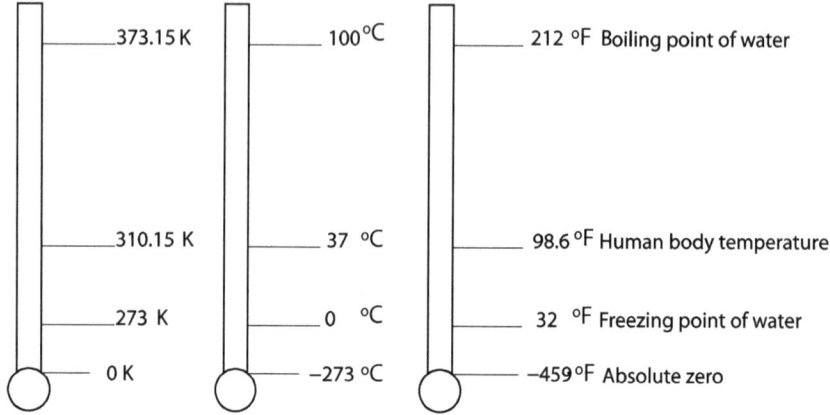

Figure 3.6. At left is a comparison of 3 temperature scales. Each functions by the transfer of kinetic energy from the substance to the glass of the thermometer and then to the liquid in the thermometer.

Kelvin, Celsius and Fahrenheit Temperature Scales

The degree *kelvin* (K) is the basic unit of temperature in the SI system. The kelvin scale is based on absolute zero or the cessation of all molecular motion because there is no kinetic energy of molecules. Absolute zero occurs at 0 K (- 273.15 °C or - 459.67 °F). Absolute zero has never been achieved. The kelvin scale (K) is used in most physics laboratories. The degree **Celsius**[103] **scale** (°C) is in everyday use in most countries of the world and all laboratories. See figure 3.6.

The *mole* (mol), also called *Avogadro's number* (N_A), is a chemical unit of mass. It expresses amount. A mole is the number of atoms in 12 grams of carbon-12 (^{12}C). Using scientific notation, Avogadro's number is 6.02×10^{23} or 602,000,000,000,000,000,000,000 in ordinary numbers. One mole of any substance has a mass in grams. It is equal to its molecular or atomic mass. The mole is reproducible and can be converted to grams, molecules or atoms. It is like saying a dozen. For example, 1 mole of H_2O has 6.020×10^{23} atoms. On the periodic table, hydrogen is noted as *atomic mass unit* 1 (**amu**) and oxygen has an amu of 16. There are 2 atoms of H and 1 atom of O. Add the amu of two H atoms and the amu of 1 O atom. They total 18 amu per mole or 18 grams.

---

103 Anders Celsius (1701-1744) was a Swedish astronomer. The Celsius temperature scale was known as the centigrade scale prior to 1948.

***The ampere* (A)** *is a unit of electric current.* It is the amount of electric charge passing a point in an electric circuit per unit time. The unit is named after André-Marie Ampère (1775-1836), a French mathematician and physicist. An electric current is the flow of electrons through a conductor.

***The candela* (cd)** *is the standard unit of luminosity or brightness.* Luminosity is a measure of the amount of power that is given off by a particular source of light in a specific direction. A candle gives off the light intensity of 1 candela.

Significant Digits

How precise a measurement is depends upon the number of *significant digits* that can be achieved. A significant digit is one that contributes to a number's precision. This would exclude zeros used as placeholders or if a mathematical calculation is far beyond the precision of the instrument. In figure 3.3, precision is only to the level of the nearest centimeter. In figure 3.4, the level of precision is to the nearest millimeter.

If a ruler is made for a microscope it is called a ***micrometer***. A micrometer may show 1 millimeter in the field of view. Each millimeter is divided into 10 equal divisions. Each division is one tenth of a millimeter. Each tenth of a millimeter may further be divide into tenths.

Scientific Notation

Scientific notation *is a way of representing numbers that are too small or too large to be written easily.* Numbers are written in the format: ***a* x 10^b** using scientific notation. The lowercase ***"a"*** represents a number that is greater than or equal to 1, but less than 10 ($1 \le a < 10$). For example, 53,000 is 5.3 x 10^4 using scientific notation. To create 53,000 in scientific notation, count the number of places from the last 0 to land between 5 and 3. This should be four places. Therefore, 53,000 is written 5.3 x 10^4. See table 3.2.

Table 3.2. Examples of scientific notation

Very Large Numbers		Very Small Numbers	
Standard Form	Scientific Notation	Standard Form	Scientific Notation
30,000,000	3x10^7	0.003	3x10^{-3}
1,000,000	1x10^6	0.000001	1x10^{-6}
13,060,000,000	1.306x10^{10}	0.00037	3.7x10^{-4}

Chapter 4 Introduction to The Science of Physics
Galileo

Galileo Galilei[104] is the most famous and influential scientist in all human history. He gave us the modern scientific method, laid down the experimental foundation of science, made major contributions to a broad number of scientific disciplines and founded an entirely new science – *physics*. Galileo gave us the rules for good science. The most important rule is to support experimentation with mathematical proof. His focus was on observation, experimentation and measurement. Before we proceed any further, some basic concepts should be explained.

Basic Concepts in Physics
1. Matter

This is an extremely basic concept. ***Matter*** *is anything that has mass and takes up space*. Matter consists of particles we know to be atoms. It may be elements or compounds and can occur in four fundamental states: solid, liquid, gas or plasma. The space matter takes up is its volume.

2. Mass

Mass *is the amount of matter contained in an object*. It is also defined as an object's resistance to **acceleration**.[105] Mass is measured in multiples or divisions of the kilogram. A rock resting on the ground will have a resistance to being moved. This is its ***inertia***.

3. Length

Length *is defined as the distance from one location (A) to another location (B)*. Multiples and divisions of the meter are used to measure length. The meter is the base unit of length in the SI system and is the agreed upon standard. It is one ten-millionth of the distance from the North Pole to the equator. We could have agreed upon a different standard, and my civilizations have in the past. The U.S. still uses inch, foot, and **mile**.[106]

---

104   Galileo was the central figure of the scientific revolution (1500-1600s). He has been called the "Father of Modern Observational Astronomy", the "Father of Modern Physics", the "Father of Science", and "the Father of Modern Science". Stephen Hawking says, "Galileo, perhaps more than any other single person, was responsible for the birth of modern science." "Galileo ... is the father of modern physics—indeed of modern science."—Albert Einstein.
105   Acceleration is increase of speed (v) per unit of time.
106   These are units used by the Romans for military and commercial uses. These units were passes on to the English and thence to the U.S.

4. Area

Understanding *area* is vital for the understanding of pressure, stress and heat conduction. *Area is the size of a surface or the number of square units that covers a surface.* The square units can be in inches (in), centimeters (cm) or decimeters (dm). Most students learn that area of a regular object is calculated by the formula length (L) times width (W). A simple example would be 1 foot square tiles on a floor in a room. Counting the tiles and adding them up will give the area in square feet. The area is the inside shape or space measured in square units. Area is also used to interpret velocity change under a graph showing acceleration of an object

5. Volume

Once the operational concept of counting squares is mastered for area, the operational definition of volume of counting cubes follows easily. Think of squares on the floor of a room. Count the squares and arrive at the area of the room. The same is true for counting the cubes to get the volume of room. Objects that take up space have *volume.* Volume is a quantity. *Volume is the capacity of a container.* It is a derived unit because it is constructed from another unit, in this case length. Construction of a cube 1 meter by 1 meter by 1 meter equals 1 cubic meter (m^3).

6. Units and Standards

There are seven agreed upon base units in the modern metric system. Each base unit or standard is represented by the quantity 1. The seven base units of the SI system are the: **meter (m)**, **kilogram (kg)**, **second (s)**, **ampere (A)**, **kelvin (K)**, **candela (cd)** and the **mole (mol)**. Multiples and fractions of base units are preceded by a prefix.

7. Ratios and Division

A *ratio is a comparison of two quantities often expressed as a fraction, a:b or a is to b.* Therefore, ratios may be written: *a/b, a:b* or written out as a is to b.

8. Vertical and Horizontal

Vertical commonly defined as something that is **perpendicular**[107] to the ground. However, if a person is climbing a hill, where is vertical? In this case, vertical can be determined by using a plumb **bob.**[108] Gravity makes the plumb bob to a horizontal plane. A horizontal plane is formed by constructing a right angle to a vertical line. This is a concept that is only good at a given location.

---

107 If a vertical line intersects a horizontal line at a right angle (90°), the vertical line is said to be perpendicular to the horizontal line.
108 A plumb bob is a weight suspended from a string.

9. Kinetic Theory of Matter

All atoms and molecules, whether they be solids, liquids or gases are in constant motion. *Kinetic energy* is the energy that an object has *because* of its motion. The faster atoms or molecules move, the further apart they will be spaced.

10. Gravity

Galileo developed the modern concept of gravitation theory in the late 1600s. His experiments proved that the Earth's gravity exerts a *"pull"* equally on all bodies. Large or small masses fall at a rate of 9.8 meters per second per second (m/s^2).

11. Weight

Weight is a measure of the force of the pull of gravity on a mass. The bigger the object, the more it will be pulled by gravity. Weight can change slightly with location on the surface of the Earth, mass does not change. If a 100 kg mass is brought to the Moon, it will have a mass of 1 kg. One kg on Earth has a weight of 2.2 pounds (lb). Its weight will be $1/6^{th}$ of that on Earth compared to being on the surface of the Moon.

12. Writing and Solving Algebraic Equations

Algebraic *equations* are number sentences with an equal sign separating two or more expressions. They are in the form A = B. A and B are expressions. The equal sign denotes an equality between the expression on the right and the expression on the left. It is a problem to be solved.

Letters represent numbers in an equation. An algebraic equation is a formula that includes one or more **variables**.[109] A variable is represented by a symbol such as *a, b, c* or *x, y or z*. These symbols represent numbers.

An equation expresses relationships between quantities. There are known quantities and unknown quantities within an equation. Unknown quantities are represented, by general agreement, by the last letters of the alphabet, *x, y or z*.

In the equation $x + a = b$, the letter x is a **variable**,[110] and a and b are constants. For example, $x + 7 = 10$ is an equation that says that some unknown, x, when added (+) to the known number 7 (the constant) is equal (=) to the quantity 10.

---

109   A variable in experimental science is a characteristic, such as acceleration, taller growth of plants, longer lived animals, expression of a disease.
110   In mathematics a variable is a value that can change by performing operations such as multiplication, division, subtraction or addition.

13. Density

Density is a physical property. Density is the ratio of the mass of a substance to its volume. This is mass per unit of volume. Density of water is measured in kilograms per liter or kg/L. One liter of pure water at 4° C has a mass of 1 kg. For example, if 1 L of water is massed on a balance and the 1 L is divided the mass by the volume. It will yield a density of water of 1 kg/L. If one masses 1 cm³, it will mass out to be 1 g. The density of water will be calculated to be 1 g/cm³

$$Density = \frac{mass}{Volume}$$

14. Specific Gravity

Specific gravity is the comparison of the density of different materials to the density of water. is 1.0 (1 kg/liter = 1 g/cm³ = 1 g/ml = 1,000 kg/m³). All are a density of 1 for pure water at 4° C under 1 atm of pressure.

15. Energy

Energy is defined as the ability to do work. Work to the scientist is the transfer of energy as a result of motion. It can be calculated by:

Work = Force multiplied by Distance (W= F x D)

Energy is classified as either potential energy (E_p) or kinetic energy (E_k). Potential energy is the energy of position. A stationary book high on a bookshelf has potential energy because it has the **potential**[111] to fall or move to a lower position. A book falling off the shelf has energy of motion or E_k. As the book is picked up and placed back on the top shelf, the book gains potential energy as it is lifted. See figure 6.1.

16. Forms of Energy

1. Mechanical energy
2. Thermal energy
   When *thermal energy* moves from one body to another body, it is called heat.
3. Sound energy
4. Electrical energy
6. Electromagnetic energy
7. Chemical energy
8. Nuclear energy
9. Elastic energy
10. Magnetic energy

---

[111] Potential: having the possibility.

17. Machines

Machines are tools that change the direction or the **magnitude**[112] of a force. A machine produces a mechanical advantage or in some cases a mechanical disadvantage. They make work easier. Machines may be *simple* or *compound*. There are six simple machines: *lever*, *wheel and axle*, *pulley*, *inclined plane*, *wedge* and the *screw*. See figure 7.1.

Galileo was the first to discover that simple machines do not produce energy, rather they only *transform* energy. Any form of energy may be transformed into another form. For example, all types of potential energy are converted into kinetic energy when objects are allowed to move to a different position. If object is at the bottom of an inclined plane, its potential energy is increased by raising it up the inclined plane. This is because it is moving to a higher position and has the potential to move to a lower position, releasing kinetic energy.

Compound machines are a combination of two or more simple machines. A wheelbarrow is a compound machine made up of a wheel and axle and a lever. An automobile engine is composed of many simple machines. It is made up of many levers, wheels and axles, screws, and pulleys. All work together to make an operating automobile engine.

18. Friction

Friction is a force that opposes sliding or rolling of one surface over another. The surfaces may be solids or fluids. The rules of friction were discovered by **Leonardo da Vinci (1452-1519)**.[113] Friction reduces the efficiency of machines.

19. Force

A *force* is a push or a pull that one body exerts on another. The force could be the pull of gravity or an externally applied force. If a book is dropped from a height, it will fall at the rate of 9.8 meters per second (9.8 m/s) at the surface of the Earth. This means that in the 1st second the book will fall 9.8 m. The book will fall an additional 8.9 m/s at the end of the 2nd second. Now the book has fallen 19.6 m. At the end of the 3rd second it would have fallen 29.4 m. This is uniformly accelerated motion, proposed and precisely calculated by Galileo and caused by the pull of the Earth's gravity. Today it is known that the pull of gravity will be less the further one is from the surface of the Earth. This idea was first proposed for first time by the German mathematician Johannes Kepler (1571-1630) in his book *Somnium* ("The Dream") written in 1608. He

---

112   Magnitude means size or amount.
113   DaVinci's rules of friction were unpublished, but found in his notebooks.

described a trip from the earth to the moon and further states how the Earth's gravity will lessen as the distance from the Earth increases. He proposed that at a point in the trip the moon's gravity will be stronger than that of the Earth and the moon will attract the spacecraft to the moon. This principle was first demonstrated with the first spaceflight to the moon when Americans Neil Armstrong, Buzz Aldrin and Michael Collins landed on the moon on July 20, 1969.

If a book across a desk, the book will slowly decrease in speed and come to a stop. This is caused by the force we call *friction*. Force is measured in units called *newtons* (N). The newton is defined as the acceleration of a 1 kilogram one meter per second per second (1 kg/s/s or 1 kg/s^2).

20. Measuring Forces

There are many ways to measure forces. The simplest way to measure a force is to use a spring scale.

21. Scalars

We can describe physical quantities by their magnitude alone. Time, mass, volume and length, distance and speed are all *scalars*. They are described by one number. A duration of 1 hour, a mass of 15 grams, a volume of 1 liter, a length of 1 meter, a distance of 2 km, and a speed of 15 miles/hour are all examples of scalars.

22. Graphing Data

There are many kinds of graphs. Line graphs are among the most commonly used graphs in science. Graphing is a method of displaying data so that is easier to visualize the relationship of one variable to another. We will use a line graph here to plot ordered pairs of data. *Graphs* have x (horizontal) and y (vertical) axes that are reference lines. We measure from these reference lines in order to find values. The x axis is almost always the *independent variable* and the y axis is almost always the *dependent variable*. For example, in figure 4.1, the velocity of a car is plotted against time by recording the velocity from the speedometer every five seconds. Velocity, can seen from the graph, is positive or negative. Time is plotted on the x axis and speed on the y axis. See figure 4.1.

Using the graph in figure 4.1, it can be seen that the automobile is undergoing constant acceleration between points A and B, but between B and C the automobile is traveling at a constant velocity with *no* acceleration. Recall why Galileo coined the term acceleration – meaning added speed. The graph shows no added speed Between points B and C. The speed is constant, or in other words there is no acceleration. Between C and D a constant deceleration can be seen.

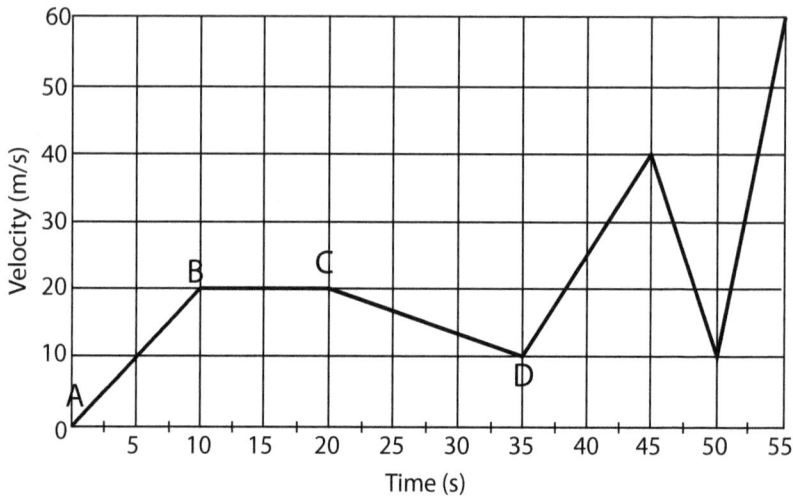

Figure 4.1. Velocity as a function of time for an automobile traveling from a local neighborhood to a nearby parkway.

Time (s)	0	5	10	15	20	25	30	35	40	45	50	55
Velocity (m/s^2)	0	10	20	20	20	17	14	10	25	40	10	60

Table 4.1. Results of a car's travel from a local street to a highway.

    The collection of all of the ordered pairs of data points for velocity (m/s) and time (s) is referred to as a *function*. Table 4.1 is all of the ordered pairs of data such as 0 and 0, 5 and 10 and so on, represent the function of velocity and time. The ordered pairs represents the functional relationship of time in seconds to velocity in miles per hour. The graph, figure 4.1, is the visual representation of all the ordered pairs of data.

Chapter 5 Motion
One-Dimensional Kinematics:[114] Motion in One Direction

Location and Distance

In order to study motion, the concepts of location and distance are necessary. Location provides a reference point. A *location* is a place or point on the surface of the Earth. The location can be assigned a marker such as the letter A. A second location can be indicated by position or point B. The separation between points A and B is defined as the *distance* from point A to point B. See figure 5.1.

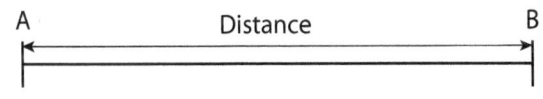

Figure 5.1. Diagrammatic representation of distance.

Speed[115] and Rate of Speed

When an object moves it is in *motion*. When something moves in a straight line from point A to point B and covers the same distance per unit of time, it is in *uniform motion*. The SI unit for *speed* (v) is meter per second (m/s). *Speed* is how fast an object is moving. *Velocity* (v) is a *vector* quantity. *Velocity* is a change in the rate of position. *Direction* must be included *with* the magnitude. *Magnitude* is a number only. **If direction is stated, use the term velocity (v).**

Speed can be calculated using any unit of length per any unit of time. Most commonly the speed of a car is measured in miles per hour (mi/h) in the U.S. and kilometers per hour (k/hr) for an in most other countries of the world. The muzzle velocity of a gun, the speed at which a projectile leaves the end of the barrel of a gun, is measured in feet per second (ft/s) or meters per second (m/s). Speed is a *scalar* quantity. Remember, a scalar quantity is described by a *single number*: a boy cleaned his room in 10 minutes or a car is traveling 50 miles an hour.

For example, a car has a speedometer that indicates its speed. Speed, or how fast something is moving, is defined as the *rate* of change in position of an object over a measured time. In this case, a car may move 10 miles in 1 hour (10 mi/h). The car can be said to have covered the distance at a rate of 10 miles for each hour. If a person walks 2 miles in one hour, the person's average speed is 2 mi/h. Rate is a *ratio* between two different quantities of the same kind; in this case distance to time. In table 4.1 for time 0 there is no increase in speed. The ratio of time to speed is 0:0, second 2 it is 5:10 and the second 3 it is 10:20 and so on.

---

114 Kinematics is a branch of classical mechanics that deals with motion of bodies.
115 Galileo was the first to measure speed as a function of distance and time.

Constant Speed and Average Speed

If a car is driven down a local street, the speed is not constant because the driver has to stop at stop signs. If a person is walking a given distance, the walk may not be constant because of stopping for traffic when crossing a road. The symbol for speed is lower case *v*.

If a car is traveling along a highway at 55 mi/h, it is possible to maintain a *constant speed* of 55 mi/h. There is no variation in speed and the speedometer will always read 55 mi/h. A speed that does not change is a constant speed. See figure 5.2. and 5.3

Most of the time we are talking about *average speed*. Average speed is the ratio of the total distance over the total time taken to travel the total distance.

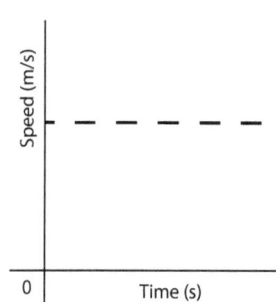

Figure 5.2. Graphical representation of constant speed.

Calculating Average Speed

Units are extremely important. To calculate average speed, distance units are divided by time units. If the distance is in meters and the time is in hours, then the units of speed are in meters per hour (m/h). If the distance is in miles and the time is in minutes (min), the units of speed (*v*) are in miles per minute (mi/min). *Average velocity = total distance/total time.*

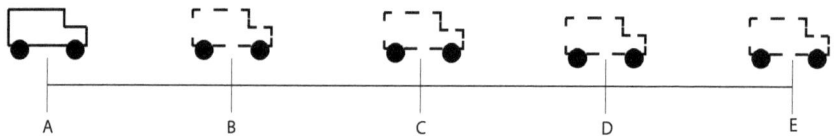

Figure 5.3. A car going at a constant speed. The car covers equal distances per unit of time.

Example:

A bicyclist travels 10 kilometers (km) in 2 hours (hr). What is the bicyclist's average speed?

Basic equation:

$$speed(v) = \frac{distance\ (d)}{time\ (t)}$$

$$v = \frac{d}{t}$$

$v = 10$ km/2 h
$v = 5$ km/h

Calculating Velocity

If a *direction* is attached to a particular speed, such as a car going *west* at 55 mi/h, then we have to use the term *velocity*. The symbol for velocity is lower case *v*. See figure 5.4.

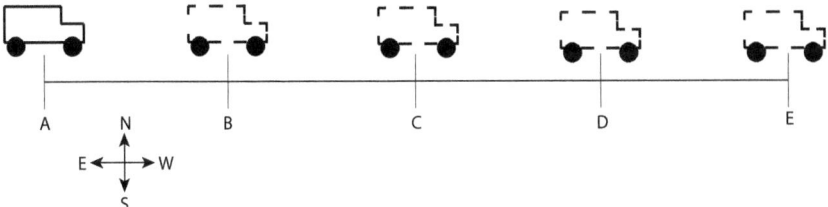

Figure 5.4. The car depicted above is traveling west at a constant velocity.

Example: A bicyclist travels **west** 10 kilometers (km) in 2 hours (hr).
Basic equation:

$$Velocity = \frac{distance}{time}$$

$$v = \frac{d}{t}$$

v = 10 km/2 hr = 5 km/hr

Note: If direction is stated in the problem, then the term velocity (v) must be used.

If two bicyclists are traveling along a road and one is traveling west and the other is traveling east, *they do not have the same velocities because they are going in different directions*. They will have the same velocities only if they are going in the same direction.

Calculating distance if velocity and time are known:
Example: A bicyclist travels **west** at a velocity of 5 km/hr for 2 hours. What distance (d) did the cyclist travel?
Basic equation:

$$v = \frac{d}{t}$$

Rearrange the basic equation to solve for d. It becomes: $d = V \times t$

$$v = \frac{d}{t} \rightarrow \frac{v}{1} = \frac{d}{t} \rightarrow (d)(1) = (v)(t) \rightarrow \boxed{d = v \times t}$$

Solution:
$d = V \times t$
d = 5 km/hr x 2 hr
d = 10 km

Acceleration
Uniform Acceleration

Acceleration of a freely falling body, such as a ball rolling down an inclined plane, is caused by gravity. In Galileo's experiment, proving that mass is independent of mass the balls regardless of mass accelerate at the same rate. Dropping a body from a height caused the body to fall to earth. Both examples illustrate **uniform acceleration**. *An increase of acceleration of a body for equal unit of time is uniform acceleration.* This kind of acceleration is due to **gravity (g)**.[116] Acceleration due to gravity is 9.8 m/s^2 and is one of the fundamental forces in the universe. It is the constant force acting on a body that is in motion. For each second of time the body is increasing its speed by an equal amount or simply stated, more speed per unit of time and that amount is 9.8 m/s for every second of time. It is a constant rate of speed. The distance the object travels increases with each second. See 5.5.

If a car is at a stop sign, it is at rest. Its speed is 0. When the driver sees it is safe to proceed, he steps on the gas pedal. After 10 seconds, the car is going 15 miles per hour (mi/h), after 20 seconds, the car is going 25 mi/h and after 30 seconds the car is going 40 mi/h. See figure 5.5. This is a *constant or uniform acceleration*. The driver keeps his foot at a constant pressure. This constant pressure on the gas pedal makes the car go at a constant speed of 40 mi/h. See figure 5.5.

Acceleration is expressed in feet per second per second or meters per second per second.

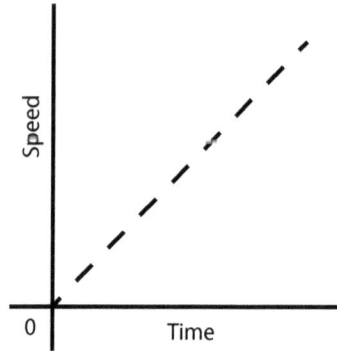

Figure 5.5. Graphical representation of uniform acceleration.

Inertia

Inertia is a property of a body that resists any change in velocity. The mass of a body is a measure of a body's inertia. The more mass a body has, the greater its inertia, and therefore, it has resistance to any change in velocity. If a small stone about 1 foot in diameter and a boulder about 6 feet in diameter are lying on the ground, it is impossible to move the boulder, but it is easy to move the small stone. The boulder has a great mass and therefore a large amount of inertia. The stone has little mass and therefore possesses little inertia. See figure 5.6.

---

116   Gravity will vary slightly depending on the location on the Earth, but 9.8 m/s^2 is the generally accepted value. It is the force of attraction of one mass on another.

Figure 5.6. The boulder at left has a great deal of inertia because of its mass. The rock at right has very little inertia because it has little mass.
Photo by Rita Anzelone

Forces Act on Bodies

A *force* is a push or a pull that one body exerts on another. The force could be the pull of gravity or some force that is applied to a body, such as pushing a box or pulling a wagon. When a force is applied to an object, the object is put in motion. It accelerates. Striking a golf ball with a golf club puts the golf ball in motion, causing it to accelerate. When the golf ball was on the tee, the ball was not in motion. When it was hit, it was put in motion. A *net force* acting on the ball causing its motion.

A force is measured units called ***newtons* (N)**. ***The newton is defined as the acceleration of one kilogram mass one meter per second per second.*** Uniformly accelerated motion of a falling body was first proposed by Galileo and is caused by the pull of the Earth's gravity. This means that each second a freely falling body will fall 9.8 m.

Balanced Forces

Forces do not always produce motion. If the forces are in balance, that is, there is no net force, then there will be no motion. Sitting on a chair, leaning on a wall are examples of forces in balance. If a person is sitting on a chair, neither the chair

Figure 5.7. Two people pulling on a rope with equal force are balanced forces.

nor the person moves because the downward force of the pull of gravity on the person is balanced by the upward force on the chair. There is no net force as a result of this action and therefore, no motion. A person leaning on a wall does not produce motion. Neither the person nor the wall moves. If two people are pulling on a rope with equal force, there is no net force because the forces are balanced and neither person moves. See figure 5.7.

Newton's Laws of Motion
Newton's First Law of Motion:
An object that is in motion will continue to remain in motion unless it is acted upon by a net force.

Case I
If a boy is pulling a wagon and his hand slips off the handle, the wagon will continue to be in motion because of inertia. No net force acts on the wagon. The wagon hits the boy in the back of the foot. The net force is the boy's foot that stops the wagon.

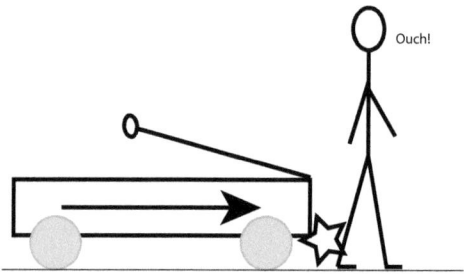

Figure 5.8. The boy loses his grip on the wagon handle and the wagon strikes the back of the boy's foot.

Case II
If a person is standing still on a stopped train and the train starts to move, the person begin to fall backwards. This is because the person's body has inertia. *Inertia* is a property of a body (mass) that resists any change in velocity. Velocity on the stationary train was 0. The person's mass is resisting the forward motion created by the train. This can be illustrated by placing a penny on a piece of cardboard and placing the cardboard on a glass of water. When the cardboard is rapidly pulled away, the penny falls into the glass of water. See figure 5.9.

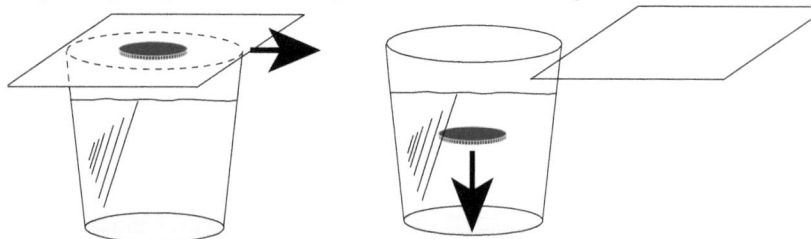

Figure 5.9. A penny resting on a piece of cardboard covering a glass of water. After the cardboard is pulled away, the penny falls into the water.

Newton's Second Law of Motion:
The acceleration of a body increases as the amount of net force applied from the outside increases. This law is expressed in the form of the equation below:

$$F = m \times a$$

F = force. Force is expressed in newtons (N). (One N = 1 kg x 1 m/s²)
m = mass
a = acceleration

Case I

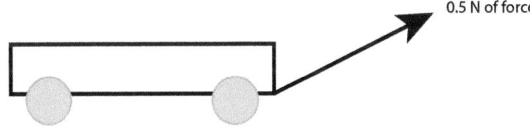

Figure 5.10. Child pulling an empty wagon.

An empty wagon is being pulled by a young girl. See figure 5.10. The harder she pulls, the faster it will accelerate. Simply put, the wagon will go faster. If the child puts more objects in the wagon, the child will have to pull harder to maintain the same rate of acceleration. So, as the load (mass) increases the force (F) that must be increased to maintain the same rate acceleration. See figure 5.11.

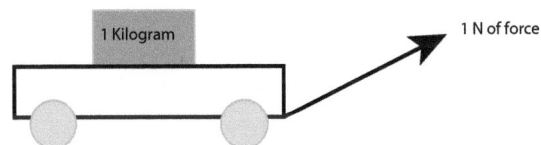

Figure 5.11. Child pulling a 1 kg weight in the same wagon.

Case II

Cars have to be more fuel efficient and still be able to accelerate at a rate that is safe enough to enter a highway safely. Gasoline powered engines have limits of force they can produce. The only practical way to increase the car's ability to increase its rate of acceleration is to decrease its mass. This is done by making cars smaller and lighter.

If the weight in the child's wagon (figure 5.11) is removed, then the child will be able to accelerate faster.

Calculating Force Given Mass and Acceleration

Problem: A hammer is swung and hits a nail. The maximum acceleration of the hammer was measured to be 3,500 m/s². The mass of the hammer is 2 kg. With what force does the hammer strike the nail?

Given: maximum acceleration of the hammer is 3,500 m/s².
Mass of the hammer is 2 kg.
Basic Equation:

$$F = m \times a$$

Find: The force (F) of the hammer striking the nail.
Solution:
F = 2 kg x 3,500 kg/m/s²
F = 7,000 kg/ m/s²
F = 7,000 N.

Calculating Acceleration Given Force and Mass.

Problem: What is the acceleration of a 2,000 kg boat that has a 12,000 N force acting on it?
Given: Mass (m) of a boat is 2,000 kg.
Force (F) acting on the boat is 12,000 N.
Basic Equation:

$$F = m \times a$$

$$\frac{F}{m} = \frac{m \times a}{m}$$

$$\frac{F}{m} = \frac{\cancel{m} \times a}{\cancel{m}}$$

$$\frac{F}{m} = a$$

Find: Acceleration (*a*) of the boat
Solution: a = 12,000 N / 2,000 kg
a = 6 N/kg
a = 6 m/s²

Acceleration Due to Gravity

Modern work on gravitational theory began with Galileo in the late 1600s. He conducted experiments proving that different masses will fall to the Earth at the same time by rolling brass balls of different masses down an inclined plane. The "falling" of the brass balls is effectively slowed down, but they are still **falling**[117] to Earth. See figure 5.12.

Balls of different masses arrived at the bottom of the inclined plane at the same time because *acceleration* was the same for a small ball as it was for a large ball. Galileo proved *mass is independent of gravity*. Gravity pulls on all objects, big or small, with the same force. This is *Galileo's Law of Falling Bodies*. Galileo calculated the pull of gravity to be very close to present measurements for acceleration of 9.8 meters per second per second (m/s/s or m/s²).

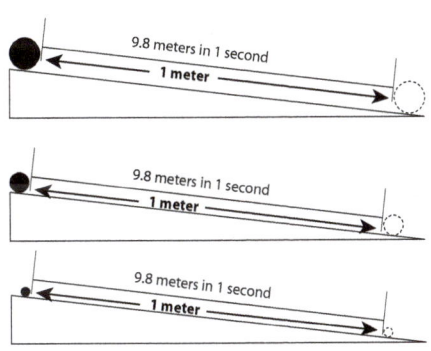

Figure 5.12. Three views of the same inclined plane showing three different masses that have been rolled down the inclined plane. Each of the three different masses accelerated at the same rate, 9.8 m/s², conclusively proving that mass is independent of gravity.

Air Resistance

Galileo hypothesized that *air resistance* played a role when lighter objects, such as a piece of paper, fall more slowly in the atmosphere than for example, a pen.

He was correct. Modern equipment allows us to see what happens if there is no air to offer resistance to a falling body.

This is demonstrated by placing a penny and a feather in a glass tube and evacuating all of the air from the tube. After quickly **inverting**,[118] the tube, both the penny and the feather will be seen to strike the bottom of the tube at the same time. See figure 5.13.

Figure 5.13. A penny and a feather dropped from position A will fall at the rate of 9.8 m/s² to position B in a vacuum.

---

117   The dropping of two balls of different weights from the Leaning Tower of Pisa, Italy is probably a myth.
118   Invert means to turn upside down.

Terminal Velocity

A few people have survived falls from great heights. Normal free fall is about 120 m/h. Documented free-falls during World War II have soldiers falling over 5,500 meters without a parachute and have survived. This is because *terminal velocity* is reached due to air resistance. The terrain of the impact area, such as snow or tree branches, also plays a role. A stewardess, pinned in an airplane, survived a fall of 10,000 meters. She suffered a broken skull, and crushed vertebrae and was in a coma for over a month, but she survived the fall. She reached terminal velocity and fell into a jungle canopy, breaking her fall.

Any body at or near the surface of the Earth that is falling under the influence of gravity only, will accelerate at the rate of 9.8 m/s². Modern equipment makes it possible to observe two bodies of different masses to be directly observed to hit the ground at the same time when dropped from the same height in the absence of air resistance. High speed cameras enable us to dissect this motion.

Galileo made air resistance almost zero by using metal balls. Air resistance has little effect on the massive balls. He knew that an inclined plane is the same as a dropping the balls vertically. Using this elegantly designed experiment, he experimentally determined with certainty that different masses fall at the same rate of speed and therefore, hit the ground at the same time. See figure 5.14.

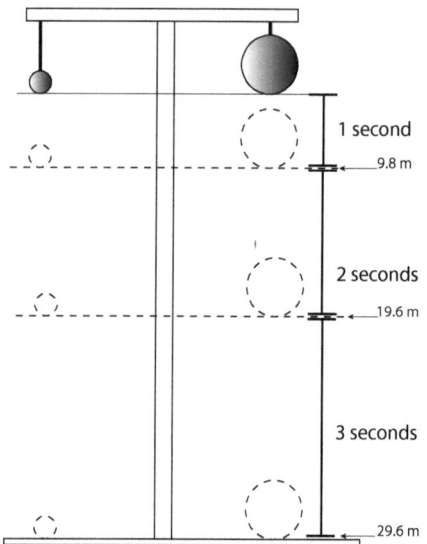

Figure 5.14. Two balls of different masses dropped from the same height fall at the same rate and hit the ground at the same time.

Two-Dimensional Kinematics: Motion in Two Directions

Projectile Motion

Galileo's was the first documented, accurate description of projectile motion. He was the first to analyze the motion of a projectile by analyzing the horizontal and vertical motion of the projectile separately. The work of Niccolo Tartaglia (1499-1557) showed that projectiles trace a curved path. Before the work of Galileo and Tartaglia, motion of a cannon ball was considered to be under the force of "impetus" where the

impetus[119] would propel the cannon ball to a point and the ball would then fall vertically to the ground. See figure 5.15. It was Galileo who showed the curve to be a parabola which could be predicted precisely and calculated mathematically.

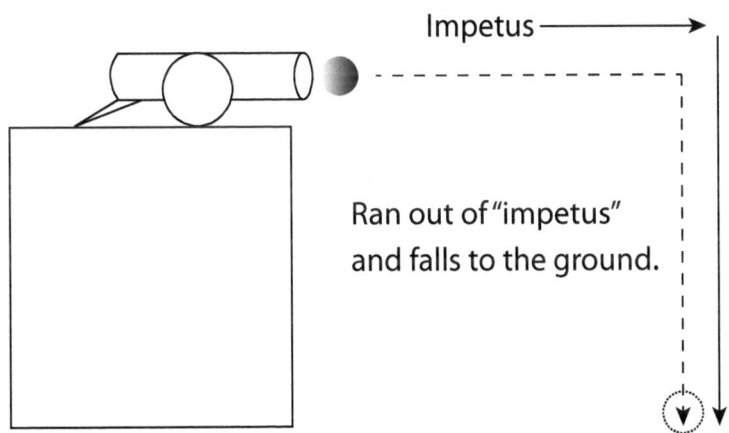

Figure 5.15. A diagrammatic representation of how people believed a cannon ball traveled in flight under the influence of "impetus."

A particle can be considered as an atom or a cannon ball. Let's consider a cannon ball fired horizontally to the ground. The forward motion imparted by an explosive charge behind the projectile is independent of the vertical motion exerted by the pull of gravity (g).

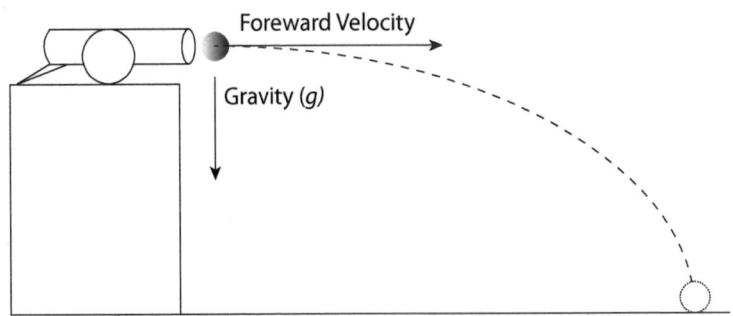

Figure 5.16. Galileo showed the curve is a parabola and could be predicted precisely and calculated mathematically.

---

119   Impetus refers to a driving force.

Motion in Circles

According to Newton's First Law, an object will continue to move in a straight line until a net force acts on it. If a boy ties a ball to a string and swings the ball in a circular path around his head, the ball will continue in the circular path. If the boy lets go of the string, the ball will move in a straight line **tangent**[120] to the circular orbit the ball is traveling.

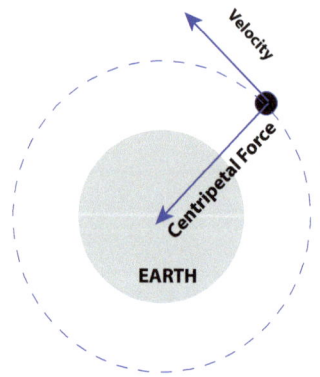

A satellite orbits the Earth in a circular orbit. Both the ball and the satellite are in uniform acceleration, but constantly changing direction because of ***centripetal force.*** Centripital force is a force that causes acceleration towards the center and causes a body to move in a curved path. See figure 5.17.

Newton's Third Law

Forces come in pairs. When one object exerts a force on a second object, the second object exerts an equal, but opposite force on the first object.

Figure 5.17. The circular orbit of a satellite in space.

Case I

A person has lifted a pail of water and is holding it in a stationary position. The person is exerting an upward force on the handle of the pail, but the pail is exerting an equal and opposite force downwards on the person's hand. While the pail of water is being held stationary, the forces are in balance. See figure 5.18. The upward force is equal to the downward force. This *results in no **net** force acting on the system (the hand and the pail).* No movement occurs. If the pail of water is lifted to be placed on a shelf, then the upward force on the handle is greater than the downward force of the pail. The forces are no longer in balance. This creates a **net** force in an upward direction, and the pail of water moves upward. See figure 5.19.

Figure 5.18. A pail of water being held in a stationary position.

Figure 5.19. Pail of water being raised to a higher position.

---

120   A term used in geometry used to describe a straight line that just touches a curved line. The straight line is sometimes simply called the tangent.

Case II

A rowboat slides backwards if a person steps off the boat and onto a dock. The rowboat will go in the opposite direction to the person's foot. If it is a large seagoing vessel, this will not happen because the vessel has an extremely large mass. See figure 5.20.

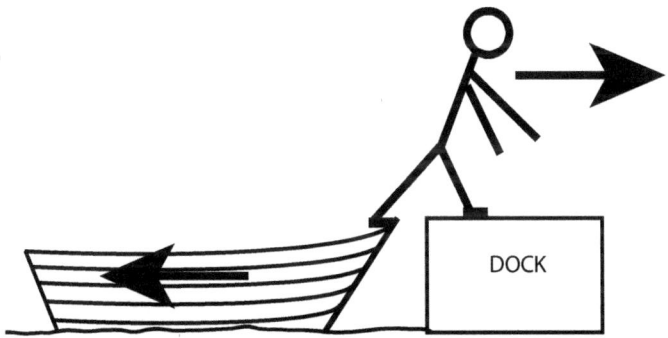

Figure 5.20. Action-reaction pair demonstrated by a person stepping off a rowboat onto a dock.

Law of Conservation of Momentum

If a boulder is at the top of a hill and is pushed over the edge, it will roll down the hill and achieve great speed. If a rock rolls down the hill it will gain speed but not as much as the boulder. A person could stop a rock, but it would be foolhardy to try to stop a rolling boulder. Clearly, *momentum* is related to the mass and velocity of an object. Momentum can be calculated by using the equation:

$$momentum = mass \times velocity$$

$$\boldsymbol{momentum = m \times v}$$

Any quantity that does not change is said to be ***conserved***. The quantity is still present. If a moving car crashes into a stationary car, the momentum of the moving car is transferred to the stationary car. The moving car will stops, but the stationary car is put in motion until it comes to rest. Total momentum remains the same. Momentum is conserved.

Another example is that of a billiard ball that is stationary on a pool table. Another ball is propelled towards it and strikes the stationary ball. The first ball stops and the second ball is put in motion. Momentum from the first ball is transferred to the second ball. Total momentum remains the same. It is conserved.

Chapter 6 Energy

Energy comes in many different forms. Energy can be transformed from one form to another form. What is energy? *Energy is defined as the ability to do work.* Energy can also produce a change in itself or anything around it.

Kinetic Energy

A boulder or a stone at the top of a hill has the potential to release energy if it rolls down the hill. This is potential energy (E_p). If the boulder rolls down the hill it also has energy of motion. This energy of motion is called **kinetic energy** (E_k). If the boulder strikes a parked automobile, it can easily produce a change in the shape of the automobile.

A moving car, a baseball in flight and a moving fan all have kinetic energy. A car moving at 20 mi/h has more kinetic energy than a car moving at 5 mi/h. However, if one of the cars is much larger than the other, the **momentum** of the large car is greater than the smaller car.

Potential Energy

A book high on a bookshelf has potential energy and no kinetic energy. If a person tries to get the book and it falls, potential energy decreases and kinetic energy increases as the book falls to the floor. See figure 6.1.

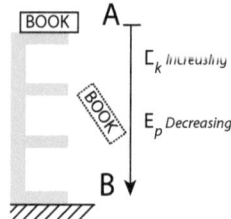

Figure 6.1. A book falling off a bookshelf is gaining kinetic energy and losing potential energy.

The gasoline in an automobile's gas tank has chemical potential energy. When gasoline is sprayed into a cylinder and explodes, the piston in the cylinder is pushed downwards. The piston is attached to a drive shaft. The drive shaft is attached to the wheels of the car. The wheels move. The energy stored in gasoline molecules becomes kinetic energy and causes work to be done. Movement of the piston is **mechanical energy**. Mechanical energy includes lifting, bending or stretching. Lifting a book from the floor, bending a sapling and stretching a spring are all examples of mechanical energy.

Work and the Transference of Energy

Moving a force through a distance is work. How is work done? Motion has to be involved in the process. Work is done when energy is transferred as a result of the motion. This is the physicist's definition, not a biologist's definition. Work can be calculated by using the equation:

$$Work = force \times distance$$
$$(W = F \times d)$$

Vertical Distance and Horizontal Distance as They Apply to Work

Carrying a book across the room is not doing work according to the physicist's definition. Lifting the book is doing work. Carrying 10 books across the room is not work, but lifting them is. *Work is only done only if an object moves in the same direction as the applied force.* For work to be done the applied force in lifting a book must be vertical. When one carries a book across a room, no work is done. This is because the applied force is in an upward direction (vertical) and the distance is the book moves is horizontal. See figure 6.2.

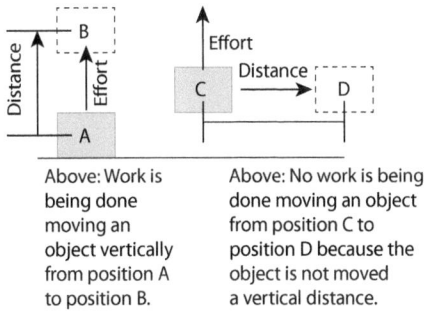

Above: Work is being done moving an object vertically from position A to position B.

Above: No work is being done moving an object from position C to position D because the object is not moved a vertical distance.

Figure 6.2. The physicists concept of work.

The Conservation of Energy

Energy can be transformed from one form to another, but cannot be created or destroyed. A pendulum bob will swing back and forth. It will reach the highest point of travel at point A (see figure 6.3) and reverse its direction at point C. It will pass through the lowest point B, where it has lowest the potential energy and highest kinetic energy. At point C, the mass of the pendulum bob has no kinetic energy, but has the greatest potential energy. As the pendulum falls back from C to B, potential energy is decreasing and kinetic energy is increasing. Maximum kinetic energy is at point B. The pendulum bob ascends to A, and as it does, kinetic energy decreases and potential energy increases until it reaches point A. Here potential energy is at its maximum and kinetic energy is at zero. If the kinetic energy and potential energy are added together at any point of travel, the sum of both will always equal zero: $E_t = E_p + E_k$.

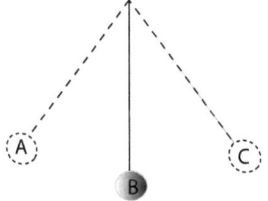

Figure 6.3. A simple pendulum.

A pendulum eventually stops because of friction at the point of suspension of the pendulum. The strands of cord suspending the pendulum bob, create thermal energy. Thermal energy is lost and is not converted back to mechanical energy. Thus the pendulum eventually stops unless an outside force is applied.

Winding a clock spring converts biological energy into stored potential energy in the spring. The potential energy in the spring is slowly released in the form of kinetic energy and the pendulum moves.

Temperature

Temperature *is defined as the average kinetic energy of molecules of a substance.* A substance with a high temperature has higher kinetic energy of its atoms or molecules than a body with a lower temperature. There are many methods of measuring temperature. They all work by reaching equilibrium with the body they come in contact. When a thermometer is put in the mouth, the glass molecules of the thermometer are bumped into by the motion of the molecules in the fluids under the tongue until the glass molecules are moving as fast as the fluids under the tongue. The glass molecules bump into the mercury molecules and make them move faster until they are moving as fast as the glass molecules. There is nowhere for the mercury molecules to go except up the hollow glass tube of the thermometer. Eventually, the whole system is in equilibrium and all the molecules and atoms of mercury are moving at the same rate. See figure 6.3.

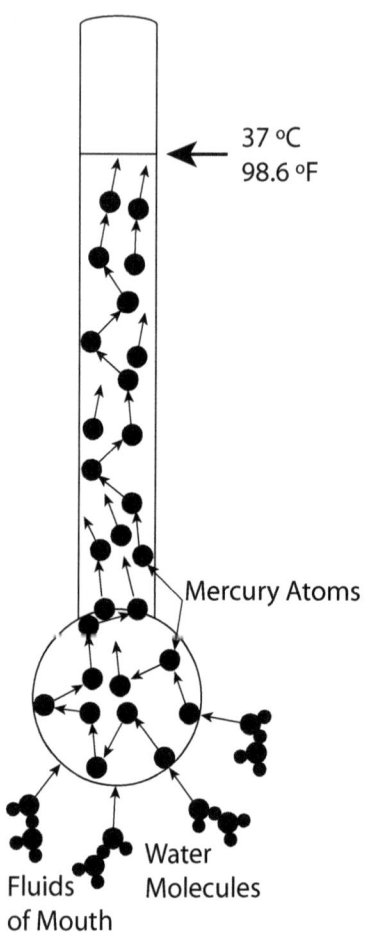

Figure 6.3. A mercury thermometer.

Thermal Energy and Heat

Thermal energy is the *total energy present in a body* caused by the random motion of the atoms or molecules in the body. If a person heats a cup of water, energy is imparted to the water by increasing the kinetic energy of the water molecules above the level of kinetic energy present in the water originally. All bodies have kinetic motion of their molecules. The only time a body would not have its atoms or molecules in motion is at absolute zero. All kinetic energy stops at absolute zero. Absolute zero is defined as 0 Kelvin, -273.15° C. or -459.67° Fahrenheit. We can come close to absolute zero, but we cannot reach it.

Chapter 7 Machines

Simple machines are constructed of only one part. There are six simple machines. See figure 7.1. A pencil sharpener and a bicycle are

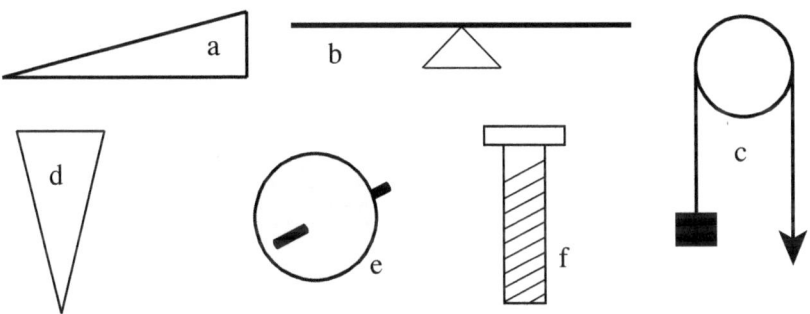

Figure 7.1. Drawings of six simple machines. (a) inclined plane, (b) lever, (c) pulley, (d) wedge, (e) wheel and axel and the (f) screw.

two examples of compound machines. Machines, in general, make it easier to do certain physical jobs by changing the direction or magnitude of a force needed to do a job. For example, opening a paint can is much easier with a screwdriver than using your fingers. The screwdriver is a lever. The lip of the paint can is the fulcrum. A downward force applied to one end of the lever multiplies the force of your hand, lifting the lid. Very little force has to be used to open the paint can.

Two forces are exerted: an effort force and a resistance force. The effort force is the force that is applied to the machine and the resistance force is developed by the machine. Machines exert the force they develop over a distance. In other words, they do *work*.

Machines make jobs easier. How do they do this? They offer us an advantage over using brute force. If a wall is being erected and concrete blocks have to be lifted straight up a distance of 3 feet, it will not take long before a person gets tired. Instead of lifting each of the blocks 3 feet straight up, an inclined plane is used to walk them up the 3 feet. Less energy has to be spent, but at the expense of time. It takes longer to walk up the inclined plane. See figure 7.2.

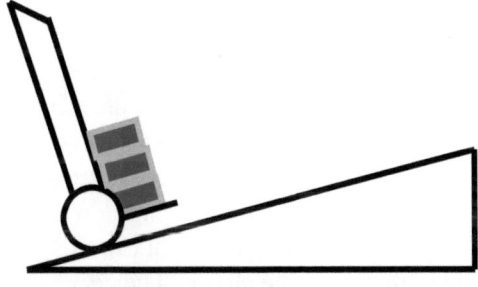

Figure 7.2. An inclined plane.

The Lever

A lever is one of the six simple machines. It is rigid bar with a fixed pivot point called a *fulcrum*. An effort is put in at one end to move a resistance at the other. See figure 7.3. F = fulcrum, R = resistance or load and E = effort.

There are three classes of levers based primarily on the placement of the fulcrum and the input and the output forces. By convention, the input force or applied (F_e) force is called the *effort* and the resistance or opposing force is called the *load* or the resistance (F_r).

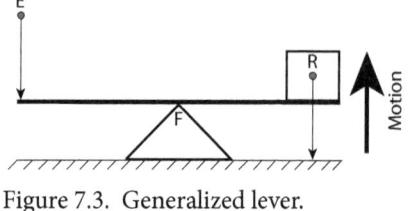

Figure 7.3. Generalized lever.

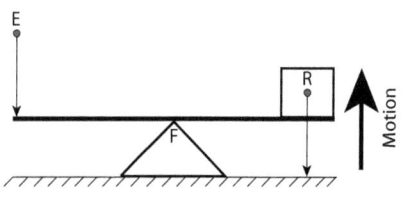

Figure 7.3a. A first class lever.

A *first class lever* has its fulcrum between the effort and resistance. The motion of the resistance is always opposite to the applied effort. A seesaw is a classic example. If a person pushes down on one side of seesaw, the other side goes up. Other examples are scissors and a claw hammer. In the human body nodding the head back and forth is an example of a first class lever. See figure 7.3a.

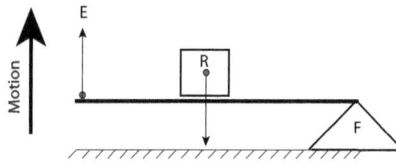

Figure 7.3b. A second class lever.

The classic example of a *second class lever* is the wheelbarrow. See figure 7.3b. In the human body, **plantar flexion**[121] of the foot or doing pushups are second class levers.

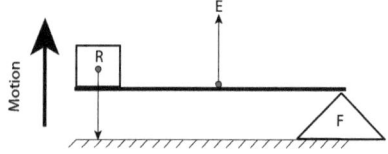

Figure 7.3c. A third class lever.

A *third class lever* has the effort between the resistance and the fulcrum. The classic example is squeezing a pair of tongs. Most levers in the human body are third class. *Third class levers always decrease effort.* They are always at a mechanical disadvantage. With the arm outstretched and a weight in the hand lifting a weight towards the body utilizes a third class lever. The fulcrum is the elbow. The effort arm of the lever is the distance from the elbow to the biceps.

---

121   Pointing the toes away from the leg.

The distance from the elbow to the weight in the hand is the effort arm. See figure 7.4.

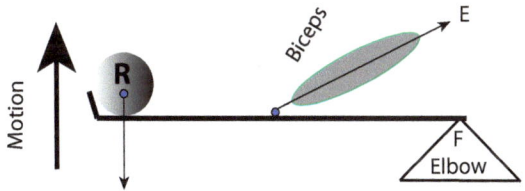

Figure 7.4. A third class lever in the human body. Raising the hand lifting a weight towards the body.

Calculating Mechanical Advantage of a Lever Given Resistance Force and Effort Force

Problem: *What is the mechanical advantage of a 6 m lever with the fulcrum located in the exact center? A resistance force (F_r) of 2 N is moved upwards by an effort force (F_e) of 2 N.*

Given: Force of effort 2 N, force of resistance of 2 N, length of lever 6 m, position of fulcrum exactly at center of lever making the effort arm 3 m and the resistance arm of the lever 3 m.
Find: The mechanical advantage of the lever.
Basic Equation:
Solution: $MA = \dfrac{resistance\ force}{effort\ force} = \dfrac{F_r}{F_e}$
$MA = F_r/F_e$
$MA = 2\ N/2\ N$
$MA = 1$

Problem: *Find the mechanical advantage of a lever with an effort arm of 4 m and a resistance are of 2 m. An effort force of 2 N is applied to one end and a resistance of 4 N is moved upwards.*

Given: Effort arm is 4 m, resistance arm is 2 m, 2 N force applied to effort arm, resistance of 4 N moves upward.
Basic equation: $MA = \dfrac{resistance\ force}{effort\ force} = \dfrac{F_r}{F_e}$

Find: MA of the lever
Solution: $MA = F_r/F_e$
$MA = 4\ N/2\ N$
$MA = 2$

Calculating Effort Force of a Lever Given Mechanical Advantage and Resistance Force

The mechanical advantage (MA) of a machine is the number of times an effort (force) is multiplied. Most of the time a machine has a mechanical advantage greater than one. If the mechanical advantage were equal to one, then there would be no obvious benefit. Mechanical advantage can be calculated by using the equation below:

$$MA = \frac{\text{resistance force}}{\text{effort force}} = \frac{F_r}{F_e}$$

Problem: *A person is using the claw end of a hammer to pull a nail out of a piece of wood. The hammer has a mechanical advantage of 15. The nail is exerting a resistance force of 3,000 N. How much of an effort force is needed to pull the nail out of the piece of wood?*

Basic Equation:

$$MA = \frac{\text{resistance force}}{\text{effort force}} = \frac{F_r}{F_e}$$

Given: MA = 15
       F_r = 3,000 N

Find: F_e

Solution: Re-arrange the equation so that F_e is on the left and the givens on the right.

$$MA = \frac{F_r}{F_e} \rightarrow \frac{MA}{1} = \frac{F_r}{F_e} \rightarrow \frac{MA \times F_e}{MA} = \frac{1 \times F_r}{MA} \rightarrow F_e = \frac{F_r}{MA}$$

$MA = F_r / F_e$
$F_e = F_r / MA$
    = 3,000 N/15
    = 200 N

Problem: *A crowbar is used to lift a heavy beam of wood off of another object. The mechanical advantage of the crowbar is 16. The beam has a resistance force of 6,400 N. What effort force (F_e) is needed to lift the beam off of the other object?*

Given: MA = F_r / F_e, rearrange the equation to solve for effort force, F_e. See above.
Find: F_e
Solution: $F_e = F_r / $MA $= 6,400$ N$/ 16 = 400$ N

Calculating Work Done

Mechanical advantage is the multiplication of a force to make work easier. The work done by simple machine can be easily calculated. Machines can increase the force applied to them, but they cannot increase the energy applied. Energy is conserved:

$$W_{out} = W_{in}$$

Machines do not create energy. Machines transfer energy. If a person hits a nail into a piece of wood, the hammer does work on the nail and the nail does work on the hammer. So then, how is work calculated for a simple machine? *Work is moving a force through a distance.* If work is put into a machine, then we designate work done as work input or W_i. Force becomes F_r or the force of the resistance and d becomes d_r or the distance the force travels. The equation may now be written as:

$$W_i = F_r \times d_r$$

Work and energy is measured in *joules*. One joule is equal to equal to 1 N x 1 m.

Problem: *How much work is done when a person applies a force of 6 N to push a box horizontally along a floor a distance of 2 m?*
Given: Force used is 6 N, 1 J is equal to 1 N x 1 m.
   Distance the force is applied through is 2 m.
Basic Equation: $W = F \times d$
Solution: $W = F \times d$
       $= 6$ N x 2 m $= 12$ J.

Problem: *How much work is done by gravity when a 2 kg book falls off a shelf a distance of 2 m?*
Basic Equation: $W = F \times d$
       $F = m \times g$
Given: $g = 9.8$ m/s^2
   $d = 2$ m
Solution: $F = m \times g = 2$ kg x 9.8 m/s$^2 = 19.6$ kg x m/s$^2 =$
   $W = F \times d = 19.6$ m/s^2 x 2 m $= 39.2$ J.

Chapter 8 Solids, Liquids and Gases
The Structure of Matter

The first to record an explanation about the nature of matter and the atom was the Greek philosopher **Democritus**. He reasoned that if a cube of any substance is cut in half and one of the halves is cut in half again and again, a point is reached where the substance cannot be cut anymore and still be the same substance. This point, the "uncuttable," is the *átomos,* or atom in English.

John Dalton (1766-1844), an English chemist and teacher, proposed an atomic theory that explained the composition of matter and how different kinds of matter react with each another. He proposed an atomic theory of matter that is very close to modern atomic theory. His atoms were solid spheres.

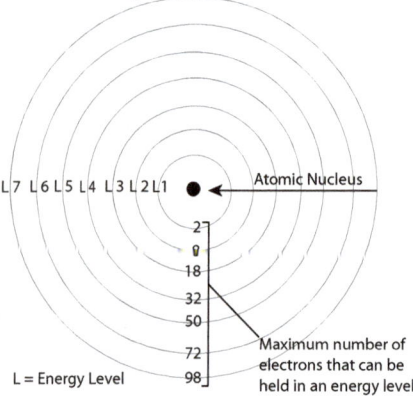

Figure 8.1. Bohr model of the atom showing energy levels 1 through 7.

Niels Bohr (1885-1962), a Danish physicist, proposed a *planetary model* of the atom in 1913, with a central nucleus consisting of protons and neutrons. His model places electrons in specific orbits called energy levels around the nucleus. Each energy level holds a maximum number of electrons. See figure 8.1. The Bohr model is called the planetary model of the atom because it resembles the planets orbiting the Sun. Bohr's model is frequently used for instructional purposes.

The Electron Cloud Model of the Atom

The *electron cloud model* is the most modern concept of the atom. It best representative how the electrons behave in an atomic structure. The electrons, although they are **discrete**[122] particles are not represented as individual particles in this model. The quantum model of the atom has replaced the Bohr model principally because it cannot be known exactly where the electrons are at any given time. We can only estimate the location of electrons within a range of mathematical probability. This model is the result of the work of many scientists. See figure 8.2.

Figure 8.2. Electron cloud model.

---

[122] Discrete means separate or distinct from other things.

States of Matter

Matter exists in four fundamental states or phases: *solid*, *liquid*, *gas* and *plasma*. Solids, liquids and gases are referred to as classical or ordinary states of matter. They exist at ordinary **temperatures**[123] and pressures. We are most familiar with solids, such as inanimate objects like tables, chairs and rocks. Liquid water is all around us and is necessary for life. Gases are more difficult to observe than solids or liquids. The gases that make up the Earth's atmosphere are colorless and odorless. Plasma requires special conditions to be achieved. It is like a gas, but the particles are ionized or carry an electrical charge. Plasma exists on the surface of the Sun and other stars and under certain conditions on Earth, such as when a fluorescent lamp is first turned on, neon signs and plasma television sets.

The faster atoms or molecules move, the further apart they will be spaced. Water molecules in the *solid state* (ice) are *moving slowly*. In the *liquid state*, water molecules are *moving faster* than in the ice state and in the *gaseous state* water molecules are *moving fastest*. As the temperature in a substance rises, its atoms or molecules move further apart. Conversely, as the temperature goes lower, the atoms or molecules move closer together. See figure 8.3, A, B and C. When a nurse takes the temperature of a patient, the thermometer is measuring the average kinetic energy of liquids and membranes it is in contact.

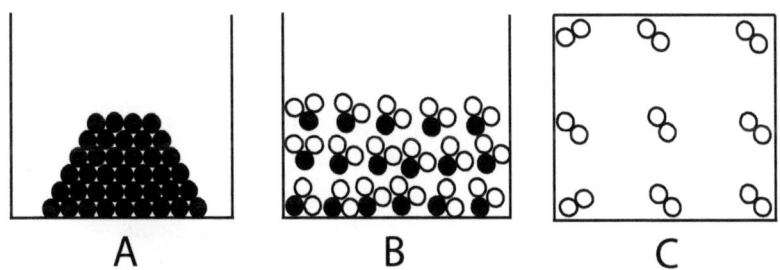

Figure 8.3. (A) Single solid spheres piled up one upon another represents a solid element. (B) A combination of two different elements all at the same level represent a molecular liquid. (C) Two of the same atoms united into a molecule (diatomic) that fills the container completely represents a gas.

Solids and liquids are not compressible. Gases are compressible. Liquids and gases take the shape of the container they are in. See figure 8.3, A, B and C.

---

123   Temperatures on Earth range from extremes of a low of about -68 ºC (90 degrees below 0 F) to a high of 57 ºC (135 degrees F). Most often we are talking about an average of 15 ºC (59 degrees F).

Kinetic Theory of Matter

The states of matter and their properties can be understood by employing the *kinetic theory* of matter. Kinetic theory states that all matter is made up of particles that are too small to be seen. In addition, the particles are in constant motion. They constantly collide with each other in a random manner because of this motion. The collisions can be observed indirectly. It is called Brownian motion.

Robert Brown discovered that if pollen grains are suspended in water and observed under the microscope, they appear to bounce around but, their path can be traced. The pollen grains are being bounced into by the water molecules. See figure 8.4.

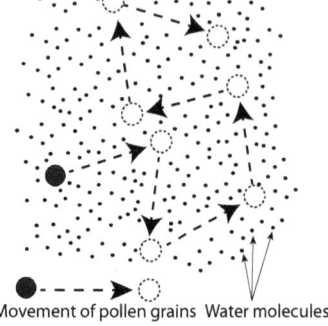

Figure 8.4. Brownian motion.

We know that there are three states of matter at ordinary temperatures: solids, liquids, and gases. These are referred to as the ordinary or classical states of matter. Plasma is a fourth state and occurs under special conditions such as at extremely high temperatures or in extremely strong electromagnetic fields.

Thermal Expansion

Adding heat or working on matter will increase the thermal energy of a body. For example, hitting a piece of iron repeatedly with a hammer will begin to heat up the piece of iron. Putting a flame under a pot will likewise increase the thermal energy of a pot. When we touch the piece of metal or the pot we say they are "hot."

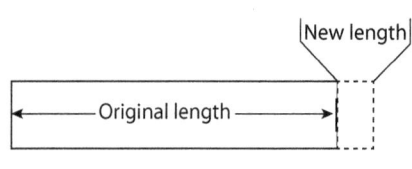

Figure 8.5. Thermal expansion.

Something can only be said to be hot if thermal energy can be transferred from one thermodynamic system to another. In this case, from a metal body to the human body and then sensed by a receptor in the skin.

As thermal energy increases, the atoms or molecules in the body bump into each other and move farther apart. As the atoms or molecules move further apart, the body expands compared to its original length. See figure 8.5. As it loses thermal energy (cools), metal contracts. Different substances expand at different rates. The expansion of a solid, in this case a metal, is a function of the type of metal and its temperature. The volume of a metal expands, not only the length. The expansion is expressed as the **coefficient of thermal expansion**.[124] The coefficient of thermal expansion for copper is higher than for iron. The higher the coefficient of thermal expansion, the greater the amount of expansion. This principle is used in thermostats and **thermocouples**.[125]

Thermostats and thermocouples consist of a bimetallic strip that utilize this principle. A bimetallic strip (see figure 8.6) is made of two different metals that are fused together such as iron and copper. Copper will expand more than iron and will "push" the iron in an upward direction. The movement of the bimetallic strip will open or close an electrical contact in response to temperature change.

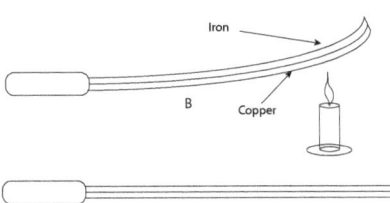

Figure 8.6. A bimetallic strip consisting of one strip of iron and another of copper fused together. Figure A is an unheated strip. Figure B shows what will happen if the strip is heated.

The Nature of a Gas

A gas is one of the four states of matter. The other three are solid, liquid and plasma. At normal temperature and pressure (**NTP**)[126] solids, liquids, gases have a characteristic amount of kinetic energy in their particles that define their states. See figure 8.7. If enough thermal energy is withdrawn from gases and liquids, for example, near absolute zero, they become solids. As thermal energy is removed, the kinetic energy of atoms or molecules of a substance slow more and more until essentially no motion of the particles exists.

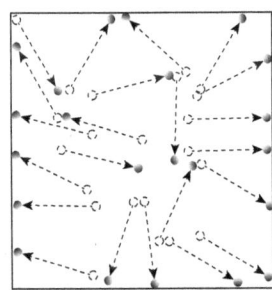

Figure 8.7. Representation of kinetic energy of gas molecules in a closed container.

Gases take the shape of the container they are in and are compressible (liquids and solids are not compressible).

Our atmosphere is composed of 1--20% molecular oxygen (O_2), 70% molecular nitrogen (N_2) and less than 1% CO_2 along with other gasses.

---

124   Change in length divided by the original length divided by the change in temperature.
125   Thermocouples are bimetallic strips that are heated and allow electricity to flow. For example, a boiler or gas heater stops running when the strip cools and a contact point opens. This is an important safety feature.
126   In the US temperature at NTP is 20 °C (68 °F) and pressure is 1 atmosphere.

Pressure

A gas exerts pressure (P) in *all directions*. If a gas is in a closed container, the gas molecules will collide with each other and with the walls of the container, thus exerting pressure on the walls of the container. A combined force of a very large number of particles, each exerting a small force against the walls of a container, is a great force. See figure 8.7. *Pressure (P) is the total force (F) exerted by the particles of a gas against the walls of the container divided by the total surface area (A) of the walls of the container.* This is expressed mathematically as: **P = F/A**. The unit of pressure is a pascal (Pa). One Pa equals 1 N/m².

Boyle's Law

Boyle's Law is one of several laws describing the behavior of a gas under varied conditions. Boyle's Law deals with the relationship between the pressure of a gas and its expansion and contraction at a constant temperature. Boyle's Law states that the pressure of a gas in a closed container varies inversely with the volume if the temperature is kept constant. See figures 8.8 and 8.9. In other words, as the volume is decreased, the pressure increases proportionately.

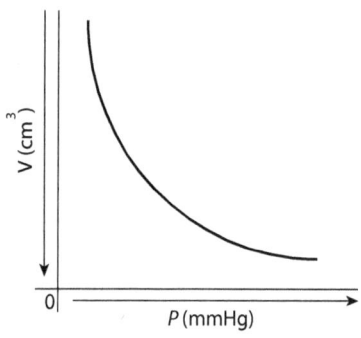
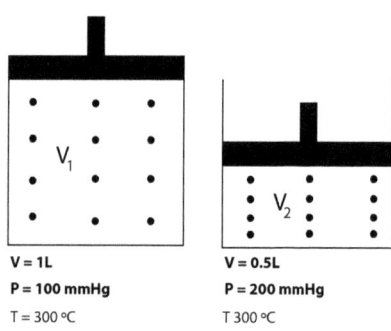

Figure 8.9. Illustration of Boyle's Law. Graph shows decreasing volume of a gas in a fixed container that creates a rise in pressure. The temperature is kept constant,

Figure 8.8. Illustration of Boyle's Law showing a fixed container with a volume of V_1. There is a movable piston at the top of the container. The piston is pressed halfway down compressing the to V_2. The temperature is kept constant,

Charles' Law

Charles' Law *is a gas law that states that the volume of a gas varies directly with the temperature if the pressure is kept constant.* If the temperature of a gas in a container that can expand is raised, the volume of the space the gas occupies increases and the container expands. See figure 8.10 and 11. An example of Charles' Law is demonstrated when an automobile tire loses volume when the outside temperature changes from a very warm day to a very cold day. As the temperature drops, the molecules of air in the tire, an expandable container, cool and their kinetic energy is decreased. As a result, less pressure is exerted against the walls of the container and the container (tire) loses volume. The operation of a gasoline-powered automobile engine is another example of Charles' Law. See figure 8.10.

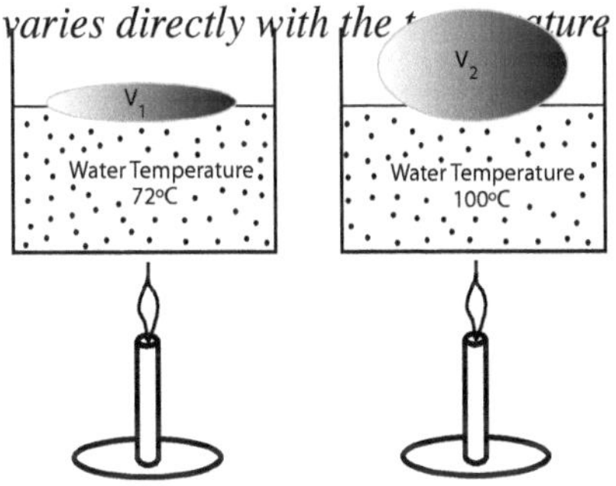

A football is inflated in a warm locker room When it is taken outdoors on a cold day it loses volume.

Figure 8.10. Experimental demonstration of Charles' law. As the air in the balloon is heated the volume expands proportionately. Pressure does not change.

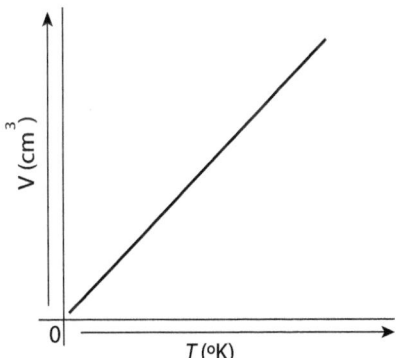

Figure 8.11. Graph illustrating Charles' Law. Volume of a gas increases proportionately with temperature if pressure is kept constant.

Fluids

Fluids are anything that flow. Gases and liquids are both fluids. Oxygen, nitrogen, carbon dioxide and carbon monoxide are examples of gases. Oxygen (O_2), nitrogen (N_2) and carbon dioxide (CO_2) are natural components of the **atmosphere**.[127] **Carbon monoxide (CO)**[128] is a gas that is a dangerous pollutant in the atmosphere. If a heating system is in the basement of a house and is not completely burning the fuel, CO will be produced and fill the basement. The gas will continue to fill the basement and flow onto the

---

127   The atmosphere is made up of 20% oxygen, 79% nitrogen and 1% carbon dioxide and other gases.
128   It is especially dangerous in the home because in the blood CO will bind the site on the hemoglobin molecule that O_2 attaches to, causing death by suffocation.

first and eventually the upper floors endangering the lives of the occupants. Imagine filling the basement with water in a closed house. Continue filling the basement and watch it fill the closed house right up to the upper floors. This is how CO will flow through the house. Remember, a gas will take the shape of its container.

Liquids are more familiar examples of fluids. They, like gases, will take the shape of the container they are in, but liquids are not compressible. See figure 8.3 b and c. For this reason, liquids are used for transferring force to different automobile wheels to stop a car. Brake fluid is used in hollow tubes that deliver force to each cylinder in a wheel to stop a car. This is a hydraulic braking system. This principle will be explained further later in the chapter.

Buoyancy

Some things float in water and some sink. Most wood **floats**.[129] Rocks, iron and other metals sink, but they all feel lighter when they are under water. These objects are being acted on by the buoyant force of water. The ***buoyant force*** *is the upward force exerted on an object that is in a liquid such as water.*

For rocks and metals, if the buoyant force is less than the force of gravity, the object will sink. If the buoyant force is greater than the force of gravity, it will float. For most kinds of wood, the buoyant force is greater than the downward pull of the Earth's gravity, so they float. Some types of wood such as *lignum vitae* (iron wood) and ebony do not float. Water has a density of 1. These two types of wood have a density greater than 1 and therefore, sink.

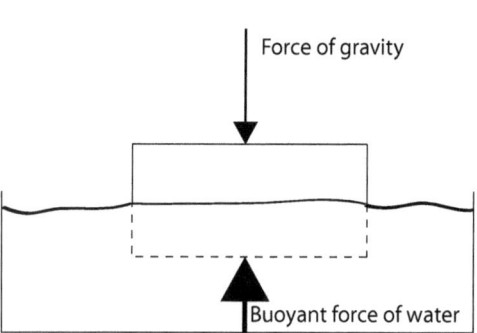

Figure 8.12. The weight of the water that the block displaces is equal to the weight of the whole block.

The ancient Greek mathematician Archimedes discovered that the *buoyant force exerted on a floating object is equal to the weight of the fluid displaced by the object.* See figure 8.12. This is known as ***Archimedes' Principle***.

Also, if a regularly shaped block of wood is placed in a graduated cylinder and it floats, the water level will rise by an amount equal to

---

129   Wood is composed of dead plant cells.

the volume of the *submerged part* of the block. This is the displacement of the block of wood. *An object that floats displaces its weight in water.*

Buoyancy can explain how a hot air balloon works. A propane tank at the bottom of the balloon heats the air inside the balloon. The molecules of air begin to move faster because of increased kinetic energy of the air molecules. The faster moving molecules exert pressure on the sides of the balloon and the balloon expands (Charles' Law). The air in the balloon is less dense that the surrounding air, causing the balloon to float in the surrounding cooler air. See figure 8.13.

Figure 8.13. At right is a balloon that is having the air inside it heat up. Notice at left, a balloon that has already risen. Photo by M. Anzelone

Pascal's Principle

Blaise Pascal (1623-1662), a French physician, discovered that if pressure is applied to a fluid. That pressure is transmitted equally in all directions. This is known as Pascal's Principle.

Fluids may be gases or liquids. Gases are compressible, but liquids are not. Whenever a person steps on the brake pedal of a car a force is exerted by the driver's foot and that force is ultimately transmitted to each wheel of the car.

A brake pedal forces a piston down into a cylinder. This cylinder, called a master cylinder, is

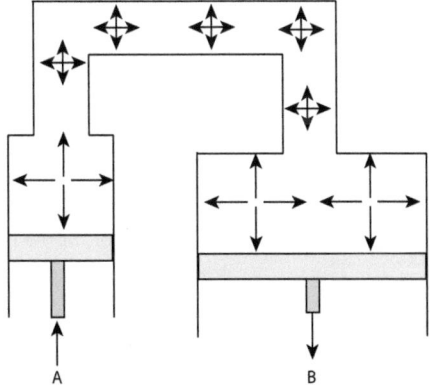

Figure 8.14. Pascal's Principle. Force is applied equally to all sides of the container, but the force is doubled in the cylinder B because cylinder B has twice the surface area of cylinder A.

filled with brake fluid (a liquid). There is a fluid in each of these wheel cylinders and all the brake lines between the master cylinder and the wheel cylinders. It is a closed system. The master cylinder pushes the brake fluid into four metal brake lines. The break fluid is pushed with equal force in all direction into the break lines, however, the only direction for the fluid to move is another movable piston in each wheel. Since liquids are not compressible, the wheel cylinders are forced outwards. When the wheel cylinders are pushed out, each cylinder causes metallic pads called break pads, to squeeze a metal disk that is fixed to the wheel. When the brake pads squeeze the metal disk, the car slows and eventually stops.

If one cylinder is twice the diameter of another, and force is equally transmitted through the fluid, then the pressure on the second cylinder is multiplied by two. This is why relatively little force exerted on a brake pedal can stop a 2,000 pound car. See figure 8.14.

Bernoulli's Principle

Archimedes' and Pascal's Principles deal with fluids that are *not in motion*. Daniel Bernoulli (1700-1782), a Swiss physicist, described the behavior of fluids (gases and liquids) that are *in motion*. His discovery of the behavior of fluids in motion explains the flight of airplanes and how a curve ball curves.

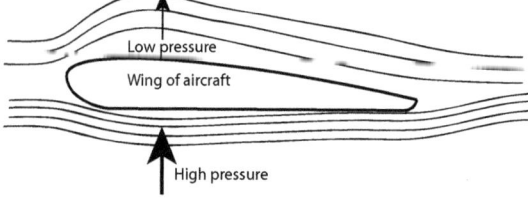

Figure 8.15. An aircraft wing illustrating the effect of Bernoulli's Principle.

When air rushes over the wing of an airplane in flight, the downward pressure of air on the upper surface of the wing is lessened. The air pressure under the wing of the airplane becomes greater relative to the upper surface causing the airplane to lift into the air. See figure 8.15.

Bernoulli was studying the relationship of the velocity of blood flow and blood pressure. He placed a straw in a small opening in a pipe through which water was flowing. He noticed that the velocity of the water was directly related to the height of the water column. Physicians, using this principle, began to measure blood pressure by puncturing patients' arteries with glass tubes that came to a sharp point.

The Italian physician Scipione Riva-Rocci invented the mercury sphygmomanometer in 1896. He employed an inflatable cuff on the upper arm to measure blood pressure, a method that is still used today.

Bernoulli and Cardiovascular Effects

A blood vessel is much like a tube in a braking system. Pressure is equal in all directions. This means that pressure is a driving force along a blood vessel and pressure is exerted laterally against the walls of the blood vessel equally as well. Think of blood flowing from one point A to another point B. It is not only the pressure (P) that drives the blood from one point to another, there is also the kinetic energy (E_k) of the flowing blood. Moving blood has mass and kinetic energy.

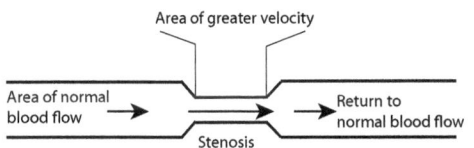

Figure 8.16. Velocity of blood in the vicinity of a stenosis.

If there is a narrowing or *stenosis* of a blood vessel and on the other side of the stenosis the normal diameter is present, the velocity of the blood increases in the area of the narrowing. Think of a lake flowing into a progressively narrowing stream. As the stream narrows, the velocity of the flowing water increases. This is due to the pressure of the water and the kinetic energy of the flowing water. See figure 8.16.

Change of State

A **physical change** occurs when matter undergoes a **change of state** or phase change. Liquids will become solids at a temperature called the **freezing point**. Solids change to liquids at the **melting point**. Liquids change state and become gases at the **boiling point**. They **vaporize**. When gases turn to liquids the process is called **condensation**. Liquids such as water do not have to boil and **vaporize** to become a gas. Puddles of water on the street dry up at any temperature above freezing. This is the process of **evaporation**.

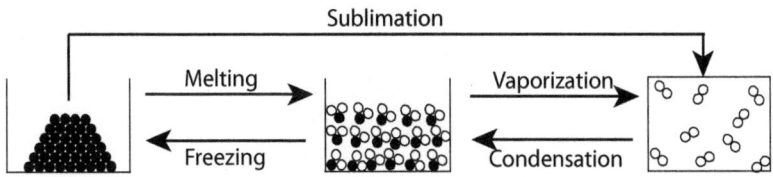

Figure 8.17. Diagrammatic representation of change of state.

Under certain conditions certain solids such as snow or solid carbon dioxide sublime. **Sublimation** is the process of going from the solid state directly to the gaseous state. The solid does not pass from the liquid state and then to the gaseous state as most solids do.

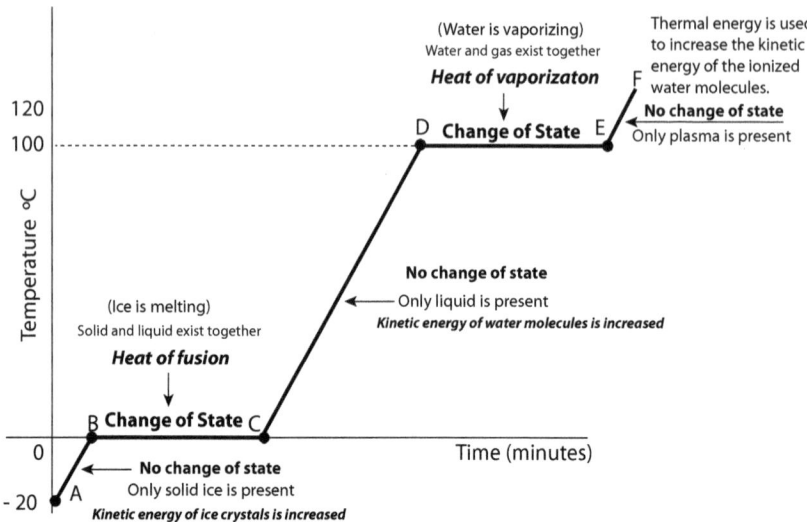

Figure 8.18. A graph showing temperature as a function of time for pure water. Thermal energy is added at a constant rate to a beaker of ice with an initial temperature of –20 °C.

Heats of Fusion and Vaporization

The graph shown in figure 8.18 represents a boiling curve for pure water. Temperature is shown on the y-axis. Time is measured at equal intervals as shown on the x-axis. Pure water will have a characteristic melting, freezing and boiling point when thermal energy is added at a constant rate at normal temperatures and **pressures**.[130] Every substance will produce a characteristic boiling curve. Boiling curves are one physical method used to identify a substances and test for a substances' purity.

Between points A and B water is in the solid state (ice). The temperature of the solid ice is rising, but not melting yet. The kinetic energy of the ice crystals is increasing. Only one phase of water exists at this time – solid.

Between points B and C ice is melting. The solid is entering the liquid state or phase. There is no further increase in temperature. The temperature remains at 0°C. Solid and liquid phases exist together. *The thermal energy entering the system is being used to break the hydrogen bonds that hold the ice crystals together.* The energy used to break the crystal structure of the ice (see figure 8.18) is called ***heat of fusion***. The heat of fusion of pure water is 334 **kJ/kg**.[131] This means that for every

130   Normal pressure is at sea level or 1 atmosphere (atm).
131   The amount of work done by applying 1 newton (*1 kg/s²*) of force through a distance of 1 meter. The joule is also a unit of energy in the SI system of measurement.

kilogram of ice, 334 **kJ**[132] is needed to melt the solid ice to water. Over six times the amount of energy is required to go from the liquid state to the gaseous state compared to the amount of thermal energy needed to melt the solid to liquid water.

Between points C and D the temperature of the liquid water begins to rise. All of the ice has melted. The thermal energy being added is increasing the kinetic energy of the water molecules. At point D, the boiling point is reached. The temperature does not increase any further. It remains a constant 100 °C. Water is changing to a gaseous state at this temperature. The liquid state and gaseous phases exist together. *The thermal energy entering the system is breaking the hydrogen bonds that hold one water molecule to another.* This energy is called **heat of vaporization**. The heat of vaporization of pure water is 2,257 kJ/kg. This means that 2,257 kJ of thermal energy is needed to convert one kilogram of water to steam or the gaseous state. After all of the liquid water has been changed to a gas, the temperature continues to rise between points E and F. This is because the kinetic energy of the gas molecules continues to increase with the added thermal energy. If the temperature continues to rise high enough, the water may become *plasma*. Intact water molecules no longer exist in the plasma state. Plasma consists of charged positive or negative ions.

Plasma

Plasma is a fourth state of matter. In the plasma state positive ions and negative ions are created that cancel each other out to achieve an electrically neutral state.

Lightning is plasma when it is discharged near the surface the Earth. The temperature of this plasma is estimated to reach 28,000 Kelvin (49940 °F).

Gases in neon light and fluorescent light tube, the upper atmosphere (the ionosphere), stars including the Sun, and tails of comets are all plasma.

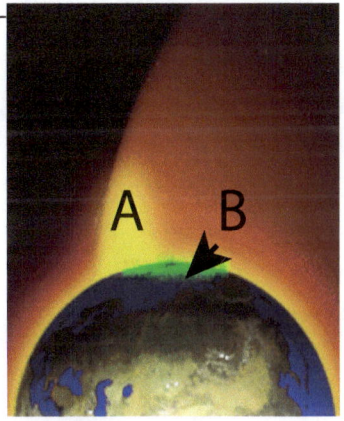

An artist's depiction of the plasma fountain caused by hydrogen, helium and oxygen ions flowing into space from the Earth's magnetic poles. The cone shape area near the north pole (A) represents ionized gasses lost to space. The round flat area (B) is the aurora borealis caused by energy returning to Earth from space. Image is a work of the Federal Government and is in the public domain.

---

[132] Kilojoules.

Chapter 9 Waves and Sound

Characteristics of a Wave

There are two kinds of waves, *transverse* and *longitudinal* waves. Compare them in figure 9.1. Sound waves are known as compressional or longitudinal waves. Sound waves are a form of *mechanical waves* such as ocean waves because a medium is needed to transmit sound. Light and other electromagnetic forms of radiation do not need a medium to be transmitted. Light can come to the Earth from the Sun through largely empty space.

The Nature of Sound

Sound waves are *mechanical waves*. Mechanical waves require a medium such as a solid, liquid or a gas for transmission from one place to another. Sound cannot be transmitted in a vacuum. We hear sound because a disturbance is caused in the air. The disturbance consists of *compressions* and *rarefactions* of the air molecules. The disturbance causes particles to move close together (compression) and then further apart (rarefaction). This is why they are also called *compressional waves*. Sound waves spread out in all directions from their source.

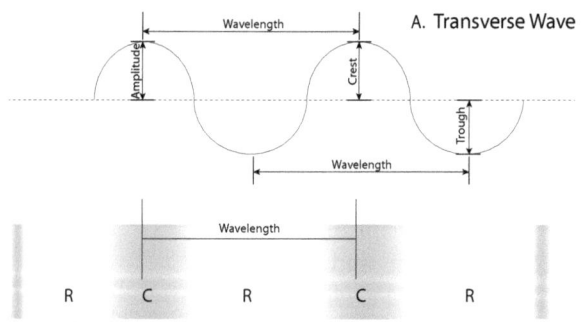

Figure 9.1. Upper portion is an electromagnetic wave. Bottom is a sound wave.

The compressions and rarefactions of sound waves move in the same direction as the wave. Compressions and rarefactions of air molecules are propagated through the air so that compressed particles move in the same direction of the wave. Compressions and rarefactions move longitudinally, hence also the name *longitudinal wave*. See figure 9.2. Light, on the other hand, is a transverse wave. The particles move vertically as the wave that carries them moves horizontally. See figure 9.1 A and 9.3.

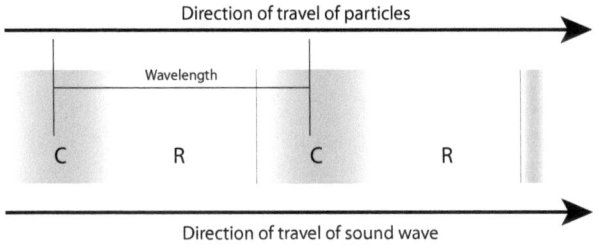

Figure 9.2. *Above*: A compressional or longitudinal wave.

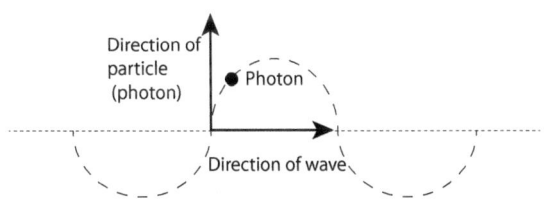

Figure 9.3. *Above*: A transverse wave.

Amplitude of a Sound Wave

Loudness of sound depends on the **amplitude** of the vibration. A loud sound is caused by tightly compressed particles of air. The more tightly compressed molecules are, the louder the sound.

Frequency of a Sound Wave

Frequency is a measure of how many times a wave passes a given point in a measured amount of time (seconds). Twenty vibrations per second mean 20 vibrations pass a given point in one second.

Frequency is related to pitch of a sound. Pitch is measured in frequencies or cycles per second (cps). The human ear can hear frequencies of 20 to 20,000 cps.

Sound does not travel in a vacuum. In a bell jar, a jar that has a bell and a battery in it, sound can be heard before the air is evacuated. As the air is evacuated, the sound of the bell becomes faint. When all the air is pumped out, the sound cannot be heard at all, See figure 9.3.

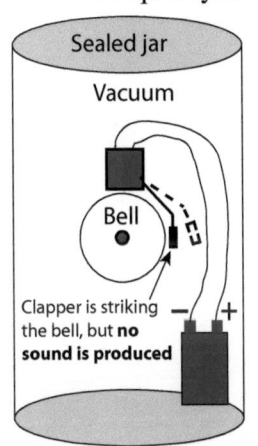

Figure 9.3. *At left*.: A sealed bell jar with a bell and battery inside.

Velocity

The **velocity** of sound in dry air at 20 °C (68 °F) is 343.2 meters per second or 768 **miles per hour**.[133] Ordinarily, when we talk about the speed of sound we are referring to the speed of sound in air, but sound travels through different materials at different speeds. The denser the material, the faster sound will travel in it. For example, sound travels faster in liquids and most solids than it does in air. In water, sound travels 1433 m/s. See table 9.1.

Table 9.1. Comparison of the speeds of sound in various substances	
Solid Material	Speed of Sound in m/s
Diamond	12,000
Aluminum	6420
Steel	6100
Glass (Pyrex)	5640
Iron	5130
Glass	3962
Hardwoods	3962
Water	1433

The Doppler Effect

The frequency of sound changes when the object producing the sound *moves relative to the listener.* A fire truck or train whistle emits a constant sound at a single frequency. As the whistle of a locomotive or siren of an emergency vehicle approaches the listener, the frequency or pitch seem to change. The pitch seems high as the vehicle or train approaches. The same pitch seems lower as it is moving away. This phenomenon is called the ***Doppler effect*** See figure 9.4.

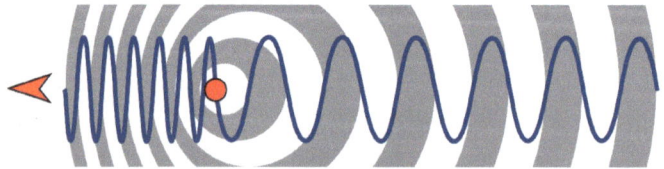

Figure 9.4. The Doppler effect. The direction of the vehicle blowing a whistle is indicated by the arrow. Air is compressed in front of the vehicle producing a higher frequency in front of the vehicle compared to the frequency behind the vehicle.

---

133   About one mile in five seconds.

Practical Application of the Uses of Sound Waves

Auscultation is listening to sounds made by the heart and lungs. A stethoscope is used to listen to the sounds produced by several organs within the body. Listening to these sounds has been used for hundreds of years to diagnose illness of the circulatory and the respiratory systems.

Sonography

Diagnostic *sonography* involves the generation of sound waves that travel through the body and are then reflected back to a detector that displays an image. The reflected image presented by sonography is usually good for viewing soft tissues such as muscle, tendons and bladders. See figure 9.5. It is commonly used for prenatal checkups: monthly observations for the first 28 weeks, every two weeks from 28 to 36 weeks and weekly until delivery. Diagnostic sonography (ultrasonography) uses sound frequencies beyond the range of normal hearing, between 2 and 18 megahertz (MHz).

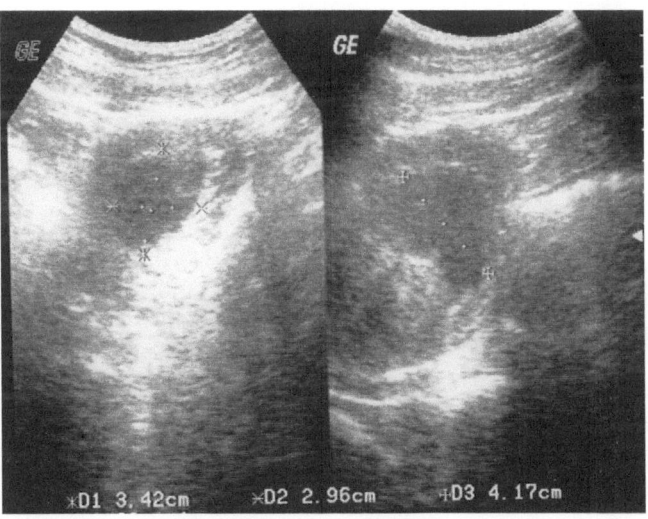

Figure 9.5. Sonogram of a human bladder.

The military use sound waves called sonar, to chart harbors and the ocean floor. Sonar an acronym for **so**und **na**vigation and **r**anging. Storms cause a shifting of the sandy bottom causing dangers to navigation. Sonar mapping of the harbor allows the placement of markers so boats can avoid hitting obstructions. Sonar is used by navy vessels to track other ships and submarines. Recreational and commercial uses of sonar include fish-finding devices.

Chapter 10 Light
Transverse Waves

Transverse waves have the properties of waves and particles. The particle, called a photon, is a packet of energy that is transferred to some object. It travels as a wave. See figure 10.1.

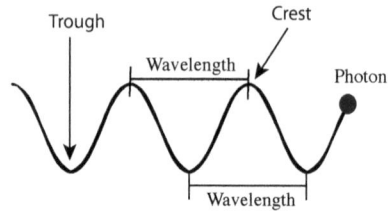

The photon is being carried by a wave that is traveling forward while the particle is moving up and down. In other words, the photon particle is traveling at a right angle to the wave direction. See figure 10.2.

Figure 10.1. Representation a photon moving as a sine wave.

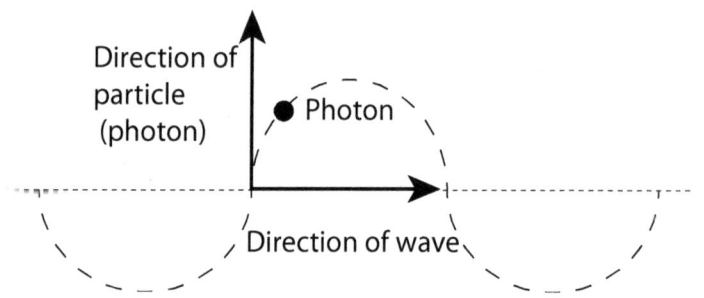

Figure 10.2. Movement of the particle and the direction of the wave.

A transverse wave oscillates up and down with regular motion. See figure 10.3. It has a regular amplitude. **Amplitude** is "how much" the magnitude of the wave oscillates, or moves up and down. The peak or crest of the wave is the highest amplitude and the trough is the lowest amplitude. The **wavelength** is the horizontal distance measured from crest to crest or trough to trough in so many seconds. This measurement represents one complete cycle. Cycles per second are also referred to as Hertz (Hz).

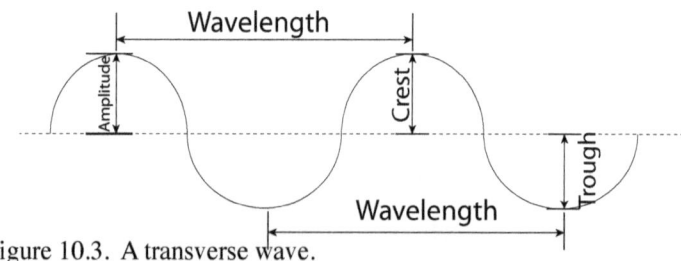

Figure 10.3. A transverse wave.

Visible light or simply light, is detected by the human eye and the eyes of other animals. Light travels in a vacuum at a speed of 300,000 kilometers per second or 186,000 miles per second. This is the fastest speed anything has been observed to move. Light is visible electromagnetic radiation.

The Electromagnetic Spectrum

We detect visible light with our eyes. Visible or white light is a small part of the electromagnetic spectrum. The ***electromagnetic spectrum*** is a way of classifying the many different kinds of transverse waves that exist. See figure 10.4. Notice that as the wavelength shortens, the frequency increases. The *frequency* of a wavelength is the number of times a wavelength passes a point in 1 second. Frequency is measured in hertz (Hz). The frequency of a wavelength determines its color.

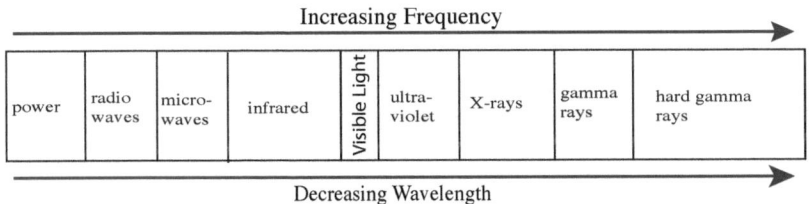

Figure 10.4. The electromagnetic spectrum.

We are surrounded by electromagnetic radiation (EMR). The term radiation is used to denote a source of energy. It has nothing to do with radioactivity. Electromagnetic radiation is energy that is given off and absorbed by particles that have an electrical charge. An electromagnetic wave has electrical and magnetic properties. Even electrical wiring in a house gives off small amounts of electromagnetic radiation. Electromagnetic radiation can travel through a medium like glass or air, but it can also travel through a vacuum like light from the Sun travels through mostly empty space to reach the Earth. Light travels at the speed of 300,000,000 meters per second. According to Albert Einstein's theory of special relativity, the speed of light is the fastest speed matter can travel in the universe.

Power and Radio Waves

Electromagnetic waves are generated by motors and power lines. Power lines are placed high on utility poles and towers to avoid the electromagnetic waves from reaching the ground. These waves are among the longest waves in the electromagnetic spectrum along with most radio waves.

Radio waves have been known since the late 1880s. They were discovered by a German physicist, Heinrich Hertz (1857-1894). Today, radio and television waves carry the energy that is received by radio and television sets. When a person "tunes into" a their favorite television or radio station, they are selecting a specific frequency.

Stars emit radio waves and these are among the longest waves, about a mile long. Most radio waves are in the range of 300 plus feet to about a foot in length.

Microwaves

Microwaves are about 3 feet to fractions of an inch in length. They carry enough energy to boil water and cook food. "Micro" refers to smaller than radio waves. The Sun radiates microwave frequencies, but the Earth's atmosphere blocks them.

Microwaves fill the universe almost uniformly. This cosmic microwave background (CMB) is thermal radiation that fills the observable universe. Cosmic microwave background is considered to be "left over" radiation from the "big bang." The "big bang" theory[134] refers to the time of the creation of the universe.

Communication for long distance telephone calling utilizes microwaves. Relay towers from the very large (see figure 10.5) to the very small help transfer as many as 5,000 conversations on one microwave channel from one tower to the next.

Figure 10.5. Microwave relay towers on Black Mountain in San Diego, California. *Photo by M. Anzelone*

---

134 The term "big bang" was coined in 1949 by Fred Hoyle, an English astronomer. In reality, he was being derisive of the theory. The theory was first proposed in 1927 by Georges Lemaître, a Belgian Roman Catholic priest

Wireless internet, Bluetooth devices and cellphones use microwaves. Shorter microwaves are used in radar. Radar can give information about speed, direction and size of another vessel on water or an airplane in flight. Radar is also used by police agencies for speed limit enforcement and is also found in garage door openers. Weather reports are greatly improved by the use of Doppler radar. Doppler radar images are generated as a result of microwaves bouncing off clouds, rain or snow.

Microwaves do not possess enough energy to ionize substances and are classified as nonionizing radiation. There is not sufficient proof that long-term exposure to microwaves, for example, holding a cell phone close to the head, has an adverse effect on health. This is quite different for high intensity exposure which can cause burns.

Infrared Radiation

The wavelengths in the *infrared* (IR) range are shorter than visible light, but longer than microwaves. Infrared waves generate thermal energy or heat. Standing in front of a fire feels warm because of the invisible infrared radiation given off.

Thermal energy from the Sun heats the Earth. Much of this infrared radiation is absorbed by the Earth during the day and reradiated back into space in the evening. Pollutants in the atmosphere trap IR, much like a greenhouse, causing a gradual heating of the Earth and causing melting of the polar ice caps.

Normal body temperature for humans is 98.6 °F (37 °C). The human body radiates IR wavelengths that can be detected by special photographic film or cameras.

Visible Radiation

Visible light or white light is one small part of the electromagnetic spectrum we can see. Light will move at different speeds in different media. In a vacuum, where there are no particles of matter present, light travels at 300,000 kilometers per second (186,000 miles per second), but in air light slows down a tiny bit. In glass light waves slow down even more. The more particles in the medium, the slower light waves move.

Visible light is made up of many different wavelengths. The eye detects the visible spectrum of light, but not above the violet frequency, nor below the red frequency. The eye detects each frequency as a differ- color. These frequencies we call color can be separated by a glass prism.

Red wavelengths bend the least and violet wavelengths bend the most. As a result, when the wavelengths of light pass from air to glass, they slow down. The change in speed causes light waves to bend, some more than others, thus separating light into the colors of the spectrum. Each color we see is a reflected frequency. All the other colors are absorbed by the object except the reflected wavelength (color). A red flag is reflecting the red frequency of visible light and absorbing all the others.

Ultraviolet Radiation

Ultraviolet (UV) light is invisible, has shorter wavelengths compared to visible light, but UV carries more energy than frequencies in the visible spectrum and less energy than X-rays. Most of the UV frequencies are not ionizing radiation. Ultraviolet rays are not sensed by humans and can cause serious sunburns. Normally the Sun and a few electrical devices such as arc lamps and black lights produce ultraviolet radiation. UV rays are produced by electric trains when an electrical arc is produced between the electrified rail and the pick up device on the train.

X-Rays

X-rays are longer than ultraviolet but shorter than gamma rays. X-rays were named by Wilhelm Röntgen (1845-1923), who is given credit for their discovery. He named them X-rays because they were an unknown form of electromagnetic radiation.

X-rays are one of the earliest non-invasive methods used in diagnosis of pathologies. X-rays penetrate soft tissue, but not bone. The object to be X-rayed is placed over a detector such as photographic film. The X-ray beam is directed toward the object and produces an image on the film. The radiation that passes through soft tissue exposes the photographic film but dense objects leave a shadow because they absorb the X radiation. Bone appears white and hollow and organs like lungs appear black. The reason hollow organs appear black is because the X-rays easily penetrate soft tissue and they expose more of the photographic film.

X-rays are used to confirm a positive skin test for tuberculosis, a potentially fatal disease that attacks the lungs but can attack the kidney, spine, bone and brain. Tubercles are dense granulomas in tissue infected with tuberculosis and will show up as white patches on the lung. They develop as a result of infection by a species of bacteria called *Mycobacterium tuberculosis*. Tuberculosis affects almost 2 billion people a year.

The roentgen (R, also röntgen) is a unit of X-ray exposure. Five-hundred roentgens absorbed by a human over a period of five hours is lethal.

Low dose X-rays are used in mammography and bone density visualization. *Angiograms* utilize x-rays to image the openings of blood vessels and the heart. The opaque material called a contrast media is injected into the circulatory system. X-rays cannot penetrate the opaque material and an image is formed showing blood flow.

Computerized axial tomography (CAT), also known as computed tomography (**CT**),[135] utilizes X-rays to produce high resolution, three dimensional images of different planes of soft tissues. Sections are superimposed one upon the other by manipulating the computerized data. This process is particularly useful if the structure of interest is hidden by other structures that conventional X-rays can not visualize clearly, such as lung cancers, coronary artery disease and kidney cancers.

Gamma Rays

Gamma rays (γ) are among the shortest wavelengths and possess the greatest amount of energy. Gamma rays are given off by the nuclear decay of atoms in stars.

Colors

Color results from the human eye's ability to detect specific frequencies of light. White light is made up of six wavelengths: red, orange, yellow, green, blue and violet. See table 10.2. When we see an object as red, red wavelengths are being reflected from the object, all other wavelengths are being absorbed by the object. An object that appears black absorbs all of the wavelengths of color and an object that appears white reflects all of the wavelengths of the spectrum.

A glass prism can break up white light into all of its constituent colors. See figure 10.6. On occasion, water vapor in the air can act as a prism and will break up white light into a spectrum of colors.

Table 10.2. Wavelengths and frequencies of colors of the visible spectrum.		
Color	Wavelength in nm	Frequency in Thz
red	700 - 635	430 - 480
orange	635 - 590	480 - 510
yellow	590 - 560	510 - 540
green	560 - 490	540 - 610
blue	490 - 450	610 - 670
violet	450 - 400	670 - 750

---

[135] An x-ray based technique that produce high contrast resolution images of soft tissue.

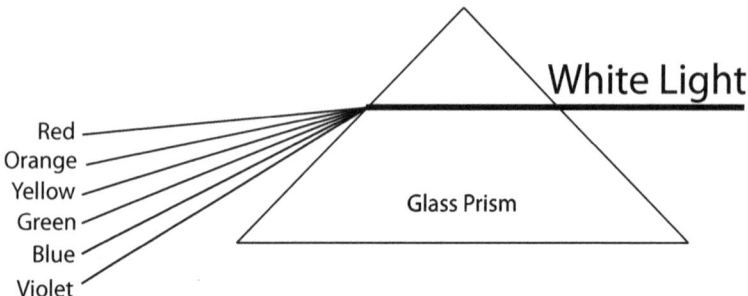

Figure 10.6. A glass prism can separate the six colors or frequencies of visible light. The six colors have specific frequencies, but the spectrum is a continuum.

Wave Properties of Light

Transmission*, *reflection*, *refraction and ***absorption*** are some of the fundamental properties of light. Transmission of light refers to the **incident**[136] light that passes through an object. There is no, or only negligible, absorption and reflection of light. Transparent material allows almost all light to pass through it. It enables a person to see what is on the other side of the transparent material. Clear glass is an example of a ***transparent*** material. See figure 10.7.

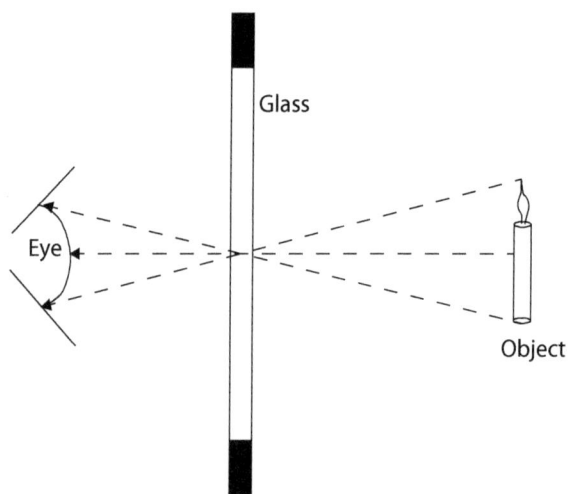

Figure 10.7. Rays of light passing through a pane of clear glass.

---

136 Incident light is a light ray that strikes a surface. A ray is a beam of light that is represented as a line.

Reflection

Reflection is a change in the direction of light waves. Light, sound and water waves exhibit this property. **Incident light**[137] strikes a reflective surface of a medium such as a mirror and its direction is changed. When light strikes a mirror-like surface, it is referred to as *specular* or regular reflection. Commonly, reflection of light occurs when a light wave returns to the source from which it originated from. The law of reflection for specular reflection is stated as: *the reflected ray will be on the opposite side of the normal and equal to the angle the incident ray makes with the normal.* The normal is an imaginary line drawn perpendicular to the mirror. See figure 10.8.

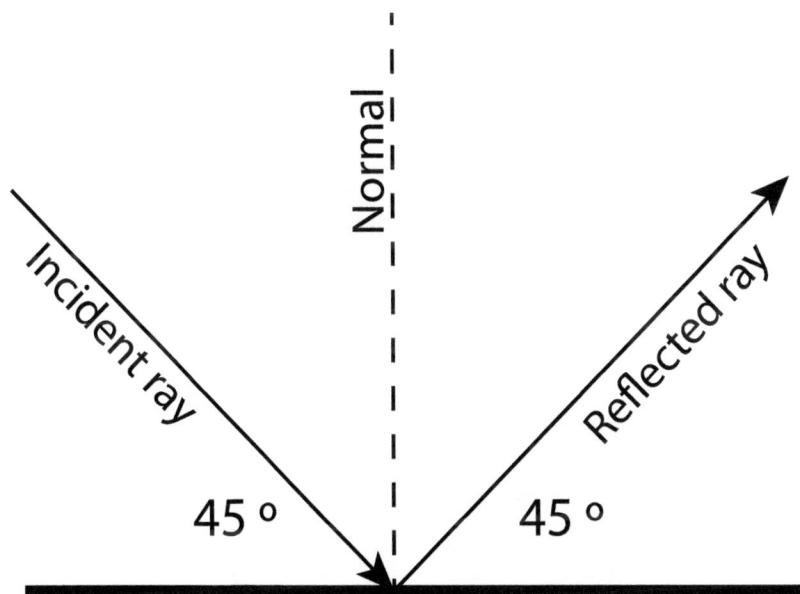

Figure 10.8. Angle of incidence equals the angle of reflection.

Refraction

A third important property of light is **refraction**. When light rays exit one medium and enter a medium with a different index of refraction, the direction of the light wave changes. This is the phenomenon of refraction. For example, the pencil in figure 10.9 seems to be slightly bent where it enters the water from the air. Both air and water are transparent media. Light changes speed because air and water have different indexes of refraction.

---
137   Incident ray is the light that approaches a mirror.

Figure 10.9. Refraction of the image of a pencil. *Photo by M. Anzelone*

Diffraction

Diffraction of light occurs when waves meet an obstacle. Electromagnetic waves, sound and water all exhibit this phenomenon. Here we will consider light only. In 1665, the Italian scientist Francesco Grimaldi was the first to record observations of diffraction, a word he coined. When a compact disc is held at an angle to the sun, light will be diffracted to reveal the colors of the spectrum. See figure 10.10.

Figure 10.10. Diffraction of sunlight by a DVD disc. *Photo by M. Anzelone.*

Chapter 11 Mirrors and Lenses
Plane Mirrors

A *flat* or *plane* reflecting surface will obey the law of reflection: the angle of incidence will equal the angle of reflection. See figure 10.8 and 11.1.

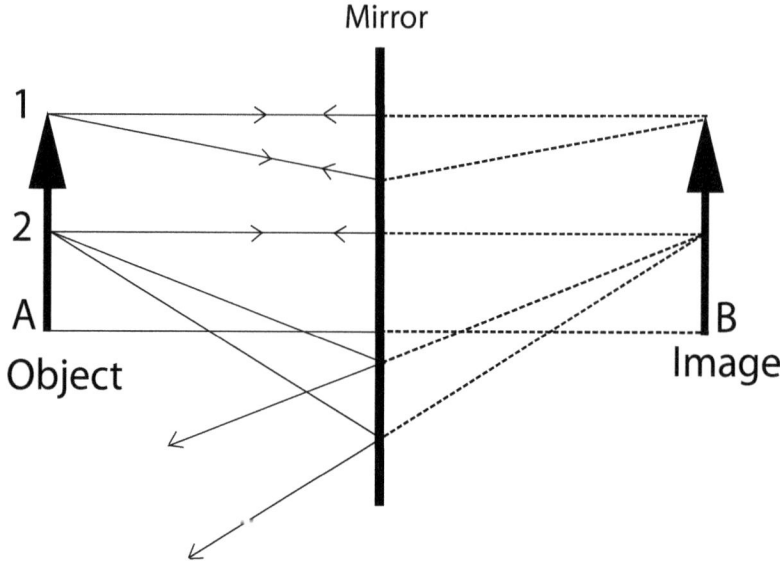

Figure 11.1. A plane mirror. Notice how each incident ray is reflected back at an equal angle.

Curved Mirrors

Curved mirrors have a curved reflective surface. Such mirrors can be curved in (*concave*) or curved out (*convex*). A commonly used curved mirror, called a *parabolic mirror*, is used in reflecting telescopes. Parabolic mirrors are effective in gather light from distant objects. See figure 11.2. Most telescopes are used for observation of objects beyond the Earth's atmosphere.

Principles of reflecting telescopes passed on to Europe in the works of the Arab scientist, Alhazen (965 - c. 1040). Building on Alhazen's writings, Galileo, Giovanni Sagredo and Cesare Caravaggi were involved in producing reflecting telescopes, but Isaac Newton built the first working model of a reflecting telescope. This kind of telescope has been come to be known as a Newtonian telescope. See figure 11.7.

Lenses

A *lens* is a curved piece of glass or some other kind of transparent material that can transmit and refract light rays. Lenses that are thicker in the middle and thinner at the edges are called *converging lenses*. These are *convex lenses*. See figure 11.2.

Lenses that are thinner in the middle and thicker at the edges are called are called *diverging lenses*. See figure 11.2. Diverging lenses are *concave lenses*.

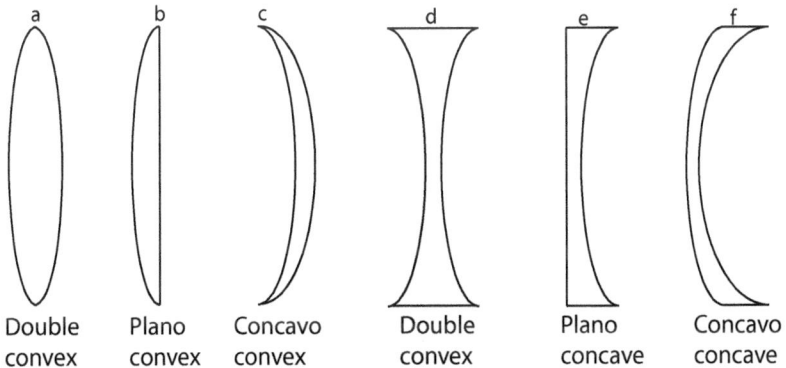

Figure 11.2. Several kinds of convex and concave lenses.

A double convex lens is an example of a simple lens. The classic example of a simple lens is a magnifying glass. One cannot draw all of the rays of light that form an image but, three main rays can be used to locate the size and location of the image that is formed. One ray enters the top of the lens and is refracted to pass through the main focal point. A second ray is not refracted and passes through the focal point. The third ray runs parallel to the center line (dashed line). See figure 11.3.

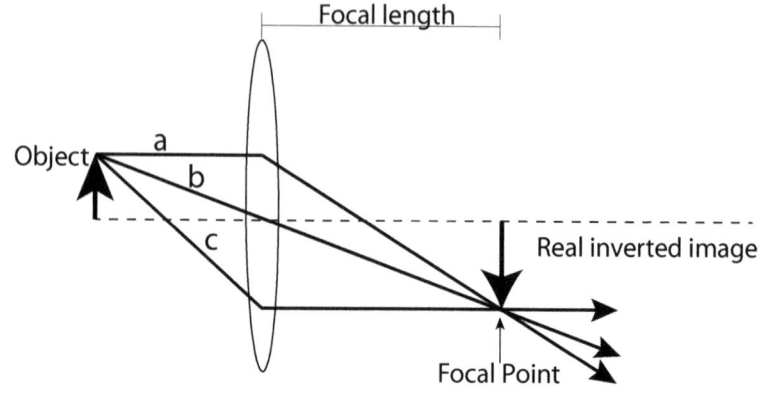

Figure 11.3. A ray diagram illustrating the path of light through a simple lens.

Vision

    Vision is a function of some organisms that allows interpretation of the surrounding environment. In humans, the eye (see figure 11.4) is responsible for detecting visible light. What is detected and interpreted by the brain is called sight or vision. The lens of the eye focuses light on the retina at the back of the eye. The retina is a **transducer**[138] that converts light energy into chemical energy and ultimately electrical energy that is transmitted to the brain via the optic nerve. The retina contains light sensitive *photoreceptors* called rods and cones. Rods and cone detect photons. The photons come into contact with a chemical substance called rhodopsin (visual purple). Rods allow vision in dim light while cones are responsible for color vision.

---

138   A transducer is a structure or instrument that converts one form of energy to another form of energy.

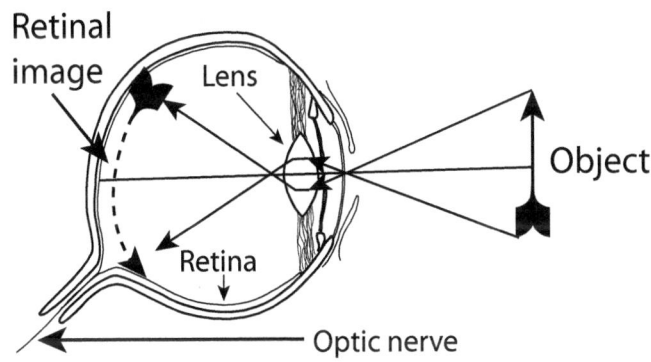

Figure 11.4. Tracing the path of light waves through the human eye.

Applications of Light Waves
Microscopes

Round pieces of glass were known to enlarge letters since Roman times. A Roman emperor is known to have used a rounded gemstone as a spectacle. However, it was not until the 13th century that an Italian, Salvino d'Armato degli Armati (1258-1312), invented spectacles by putting single lenses in frames to correct poor vision. The technological progression from seeing clearly to seeing the invisible was a logical one.

The Dutch naturalist Anton van Leeuwenhoek (1632-1723) constructed and used simple microscopes of high quality that magnified objects 100 to over 200 times. Although Leeuwenhoek lacked a formal education, he possessed great skill and curiosity. He was the first to observe and draw bacteria, protozoa, algae, yeast cells, sperm cells, and blood cells. More importantly, he published his findings even though some individuals claimed to have used simple or compound microscopes before he did. His observations of bacteria and his published written records earned him the title of "Father of Microbiology" and "Father of Bacteriology."

Robert Hooke (1635-1703), an English scientist, studied the elasticity of springs, the behavior of gases, gravity, and the refraction of light. He also constructed and used compound microscopes. He is probably best known to most students for his microscopic studies of cork, in which he described what he observed as "cells." We know today he was observing the walls of once living cells of the cork tree. Today we use "cell" to describe the living building blocks of organisms. Since the days of the early microscopists, technology has continually improved microscopes to enable these scientists to peer deeper into this invisible world.

Modern Microscopes

The eye can see objects that are at least 0.004 of an inch, or about 0.1 millimeters. *Microscopes* help extend the sense of vision below this threshold. There are many different kinds of microscopes. Each has specialized uses. We will consider some of them at a later time.

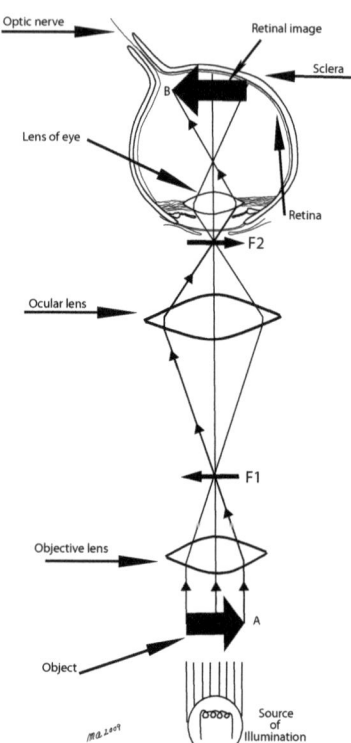

Figure 11.5. Path of light through a modern microscope's lens system.

Most microscopes use light waves and glass lenses that bend or refract the light in such a manner that the image is magnified. The best light microscopes allow magnification of a specimen up to about 2000 diameters. A poor quality microscope may magnify a specimen more, but it may not give a clear image because the lenses have poor resolving power. Resolution is the ability to distinguish between two objects clearly. For example, draw two straight lines with a ruler on a piece of paper. Draw the lines parallel to each other and 1 mm apart. Gradually move further back from the paper and you will not be able to tell if there is one line or two. Your eyes do not have the ability to resolve the lines. Glass lenses have the same limitations.

Glass lenses have certain imperfections or aberrations that have to be corrected for. One imperfection is spherical aberration. The image will be distorted at the edges of a lens because light waves are refracted more near the edges of the lens than at the center. A second aberration is chromatic aberration, which causes spikes of color to appear at the edges of the lens. This occurs because white light is made of different frequencies or colors of light. Since different frequencies of light travel at different speeds, the colors do not focus together at the edge of the lens.

These imperfections are corrected for in modern light microscopes. Objects as small as bacteria may be seen with light microscopes, but not viruses. Viruses require electron microscopes to be seen. A student should be familiar with the parts of a standard student compound light microscope that one will find in most colleges. See figure 11.5.

Telescopes

Telescope is from the Greek meaning "seeing things far away." There are two basic types of telescopes: refracting and reflecting. **Refracting telescopes** bend light using at least two lenses. One is an objective lens and a second lens, the eyepiece, magnifies this image. See figure 11.6.

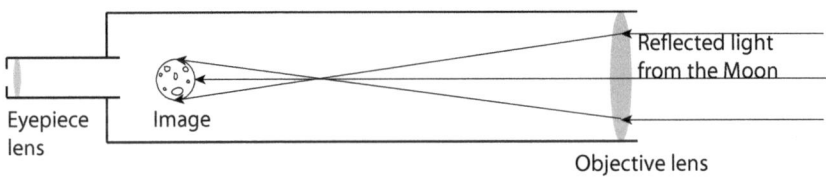

Figure 11.6. Principle and construction of a typical refracting telescope.

The idea of using mirrors to gather light was not new. Like many discoveries in science, one idea is built upon another. Galileo Galilei (1564-1642) and Giovanni Sagredo (1617-1682) had discussed this kind of telescope many years earlier, but Sir Isaac Newton (1642-1727) is credited with its invention. A **reflecting telescope**, also called a Newtonian telescope, has a primary concave mirror, parabolic in shape and a secondary plane mirror. Newton produced the first reflecting telescope that worked. If you have ever looked through a refracting telescope you will recall seeing tiny slivers of different colored light. This is chromatic aberration and is a major fault of refracting telescopes. See figure 11.7.

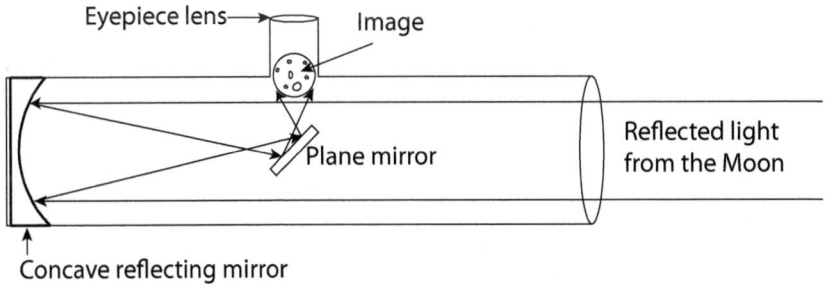

Figure 11.7. The principle and construction of the Newtonian reflecting telescope.

Polarized Light

Electromagnetic radiation radiates in all directions from a single point. A light bulb in the middle of the room shines light in all directions. See figure 11.8. Polarized light will radiate in only one direction or plane. Polarized light can be achieved by passing light radiation on many planes through a filter, basically a slit. Light will only be allowed to pass through the slit in one plane. See figure 11.9.

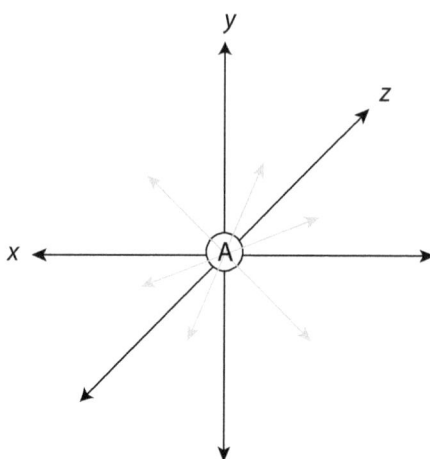

Figure 11.8. Planes of unpolarized light (x, y and z) emanating from source A.

Lasers

Laser is an abbreviation for **L**ight **A**mplification by **S**timulated **E**mission of **R**adiation. Lasers allow for very little diffraction and therefore have a very narrow, concentrated beam of a single frequency of light. In the consumer electronics field, laser technology makes possible barcode readers, laser disc readers, printers and players.

Lasers in medicine are used in surgical procedures, kidney stone treatment and vision correction. Laser may be used to treat severe acne, for tattoo removal, reduction of striae and cellulite and for hair removal.

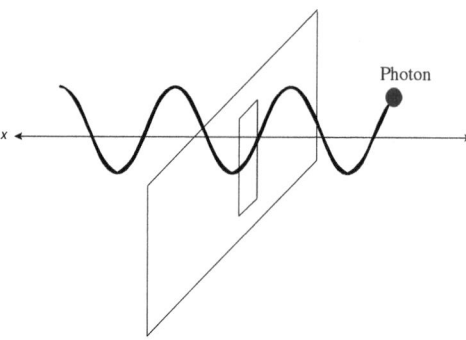

Figure 11.9. Diffraction grating allows light to pass through in one plane.

Military use of laser technology allows for target acquisition and for delivering munitions to targets (smart bombs).

Chapter 12 Electricity

Static Electricity

Static electricity is electrons at rest. Static electricity is caused by having an excess of electrons on the surface of some object. If there is an excess of electrons, they will jump or discharge to an object that has a lesser amount of electrons. Electricity in motion is termed *current electricity*. Current electricity, in contrast to static electricity, needs a conductor, usually a copper wire, through which electrons flow.

If a glass rod is rubbed with a piece of silk, electrons are transferred by friction to the silk from the glass rod, leaving the glass rod positively charged. If a hard rubber rod is rubbed with fur, electrons are transferred by friction to the rubber rod, making the rod negatively charged. This was discovered and explained by Benjamin Franklin (1706-1705) when he observed that a **pith**[139] ball, which is positively charged, was attracted to a rubber rod indicating the rubber rod is negatively charged. Conversely, the repulsion of the positively charged pith ball by the glass rod means that the glass rod had to be positively charged. See figure 12.1.

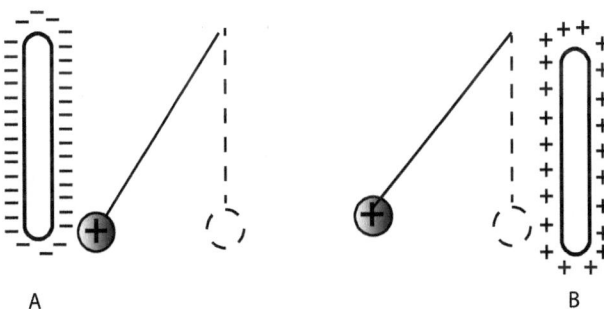

Figure 12.1. Illustration of the behavior of electrical charges in static attraction and repulsion. This was a form of early electroscope with a negatively charged rubber rod at A and a positively charged glass rod at B.

Conductors

Substances that can carry electrical current are called *conductors*. A metallic conductor has movable or mobile free electrons on its surface that move from an area of low electrical potential to an area of higher electrical potential. Most conductors are metals such as copper, aluminum, iron and silver.

---

[139] Pith is the spongy mater in the stems of plants. Rolled into a ball it can be used to detect static electrical charges.

Insulators

If a substance does not have mobile electrical charges, it is an *insulator*. Insulators are nonconducting material that resist the flow of electric current. Electrons cannot flow freely. Glass, rubber, paper and plastic have a high resistance to the flow of electrons and are good insulators. Insulators are used to separate wires in electrical equipment. Wires from utility poles that enter a home or wires that transmit electricity over long distances need to be insulated from the towers that the wires are attached. The wires that bring electricity into a home are set high on utility poles because of the dangerously high electrical current. Transmission wires are on higher towers because they carry even higher current of the magnitude of kilovolts, or thousands of volts. See figure 12.2.

Figure 12.2. These electrical transmission lines carry over 100,000 volts. Notice the series of disks strung together used to insulate the high voltage wires from the metal towers. *Photo by M. Anzelone*

Electric Current

Electric current is a flow of electrons from some object that has an excess of electrons to an object with fewer electrons. When the electrons flow in a conducting wire, a current is produced. Current is measured in **amperes** (SI unit symbol: A). Current is a measure of the number of electrons that pass a given point per unit of time. Electrons have a very tiny amount of negative electrical charge. This would not be practical to measure. For this reason, a larger unit is used to describe an electrical charge, the coulomb (C). One *coulomb* is equal to 6.28×10^{18} electrons (628,000,000,000,000,000 electrons is a lot of electrons). One coulomb of electrical charge passing a point in a conductor in one second is *1 ampere*. This is why as little as 2 amperes can kill a person!

Voltage

Voltage, a term most people are familiar with, is *electrical potential difference*. Electrical potential difference can be measured in volts or joules per coulomb. We will use volts. In order for electricity to flow, a difference between two points has to exist. It is the same as two points on land where a pond is located high on a hill. The water has a potential to flow downhill to form a current of water. It is similar to the flow of voltage. Voltage has the potential to flow from point A to point B, but only if there is a potential difference per unit of charge. This difference is called *electromotive force* (EMF) and is measured in *volts* (V or E[140]). Volts are the push behind current, much as the stored potential energy up on a hill is the stored energy of water that has the potential to flow because of a difference in elevation.

Resistance and Ohm's Law

Current moving through a conductor is moving between two points. The current is directly proportionate to the difference in potential between the two points. Let's return to the example of the pond up on a hill that has the potential to flow downhill. Let's call point A the pond location and point B the stream's end at the bottom of the hill. The higher a pond is on the hill, the greater the potential difference between points A and B. Water in the pond flowing downhill in a stream will flow with a greater force than from a pond that is lower on the hill. Voltage is like the pushing power of the water similar to the EMF of electric current.

In a metallic conductor, the intensity (I) is always directly proportional to the EMF that is applied. The greater the current, the greater

---

140   E and V are used interchangeably. They are one and the same.

the pushing power or EMF. Depending on the kind of material the conductor is made of iron vs. copper, or the thickness of the wire, resistance (R) will increase or decrease. The intensity of the current is directly proportional to the ratio of V (voltage) :I (current) and can be expressed as:

$$I = \frac{V}{R}$$

This is a mathematical equation of Ohm's Law. Georg Ohm (1789-1854) was a German physicist and mathematician. Current varies in this example and resistance is constant and does not vary with the current. As the current is increased the voltage or EMF increases proportionally. Simply stated, increasing the voltage increases the current by equal amounts. As an illustration, if a wire of the same length and diameter is connected to batteries of increasing voltage, then greater current is produced.

Electrical Circuits

An electrical circuit has a source of electromotive force and one or more complete paths for the flow of electrons. The path that electrons flow along is usually made of metallic wires. As the flow of electrons continues in a circuit, electrical potential is reduced. This is called a *voltage drop*. Series circuits and parallel circuits are two kinds of closed loops of wires that conduct electrons.

Figure 12.2. Symbol for resistance.

Series Circuits

A series circuit is one in which the electrons flow through all of the resistances that are connected one after another. An electric light bulb (or any other electrical appliance) is a resistance. Resistances in a circuits are symbolized with the symbol in figure 12.2. In this kind of circuit, if one bulb is removed the circuit is broken and all of the lights will go out because there is only one path for electrons to flow along. This is why when a circuit breaker trips to the off position, the flow of electrons to lights and appliances on that one series circuit all go out. There is only one path for electricity to flow in order to energize the lights or appliances.

Also, if three 1.5 volt lamps are connected in series, the same current is applied to each bulb, but the voltage drop is 1.5 volts for each bulb. If the battery is a 3V battery, then there will not be any voltage left to light all of the bulbs brightly. They will all be dim. See figure 12.3.

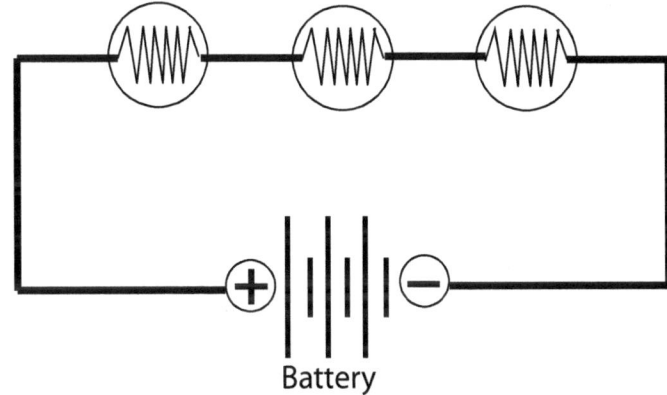

Figure 12.3. Series circuit showing three bulbs (resistances) connected in series.

Parallel Circuits

Parallel circuits have two or more paths for the flow of electrons. If we look at a parallel circuit consisting of electric light bulbs, it is obvious that if one bulb burns out, the others will stay lit because there is an alternate path for electrons to flow through to reach the other bulbs. The voltage and current are the same for each bulb. See figure 12.4.

Parallel resistance can be found in the human circulatory system. Each organ is supplied by an artery that branches off the aorta. An artery becomes a smaller vessel called an arteriole. Emerging from the same organ is a vein that has emerged from smaller venules.

In figure 12.5 two parallel vessels. An artery, brings blood to the organ and a vein drains the organ. As blood enters a vessel, the greatest resistance is at the level of the diminishing diameter of the artery. The venules connect with vessels connected in series, the capillaries. Pressure does not increase there because of the alternate paths blood can take. Any interruption to an artery stops blood flow to that organ.

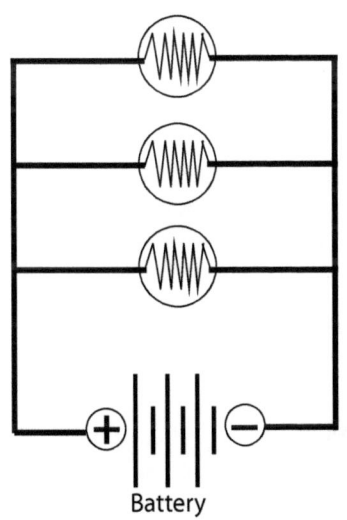

Figure 12.4. Parallel circuit.

115

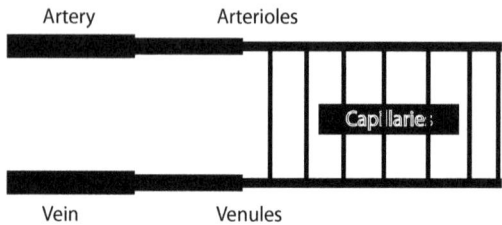

Figure 12.5. Series and parallel arrangement of blood vessels.

Electric Power

Electric power is the rate at which electric current is delivered to a home, office, or appliance. It is a measured amount of electric current flowing, measured in watts. A watt is measured in joules of energy that flow in 1 second. Power companies charge retail customers by the 1,000 watts of power or the kilowatt. The cost of power consumption of an appliance can be determined by using the following formula:

$$\text{cost} = \frac{\text{watts x hours x cost / kilowatt hour}}{1000 \text{ watts / kilowatt}}$$

Problem: *What is the cost per hour of running an electric oven that uses 1,000 watts of power for 1 hour if the cost of a kilowatt hour is 0.12 cents?*
Given: Cost = 1,000 watts used
   Cost per kilowatt hour is 0.12 cents
Basic Equation:

$$\text{cost} = \frac{\text{watts x hours x cost / kilowatt hour}}{1000 \text{ watts / kilowatt}}$$

Solutuion:
Cost / kilowatt hour =1,000 watts x 0.12 cents/ 1,000 watts/kilowatt
   = 120 cents or $1.20

Chapter 13 Magnetism
Magnets

Some substances such as iron, nickel, cobalt and mixtures like steel[141] will respond to a *magnetic field*. Iron filings sprinkled on paper with a bar magnet beneath it display the lines of force because of this attraction. See figure 13.1. A magnetic field is not visible, but can be traced (see figure 13.1) using the iron filings technique. Some materials, such as iron and steel, are strongly attracted to a magnetic field. They display *paramagnetism*. Other substances, such as another *magnet*, are repulsed by a magnetic field. This is called *diamagnetism*. Some substances are barely attracted to a magnetic field and are termed *nonmagnetic*. Copper, aluminium, gases, and plastic are non magnetic. However, if oxygen is supercooled to the solid state it exhibits magnetic properties.

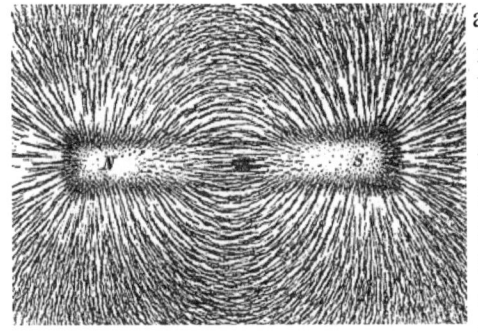

Figure 13.1. A bar magnet under a piece of paper that as sprinkled with iron filings. This image is in the public domain.

Certain substances have a permanent magnetic field caused by *ferromagnetism*. They are permanent magnets. They can be natural permanent magnets called loadstones, or they can be man made magnets, called electromagnets. Bar magnets display a south pole and a south pole. See figure 13.2. Opposite poles such as a north and a south pole will be attracted to each other. Like poles will repel each other. See figures 13.3 and 13.4.

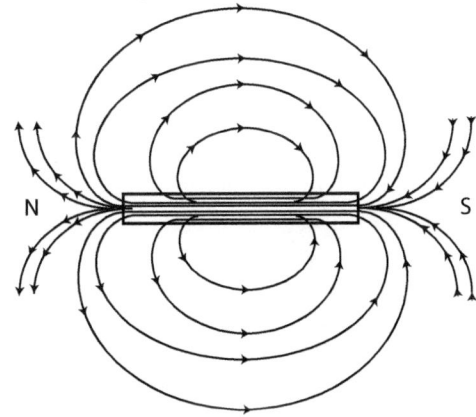

Figure 13.2. Lines of force around a bar magnet.

---

141   Steel is a mixture of solids in solids. It is type of mixture called an alloy. Most steel alloys are composed of the elements iron and carbon, but manganese, chromium, vanadium, nickel and tungsten are added for special types of steel.

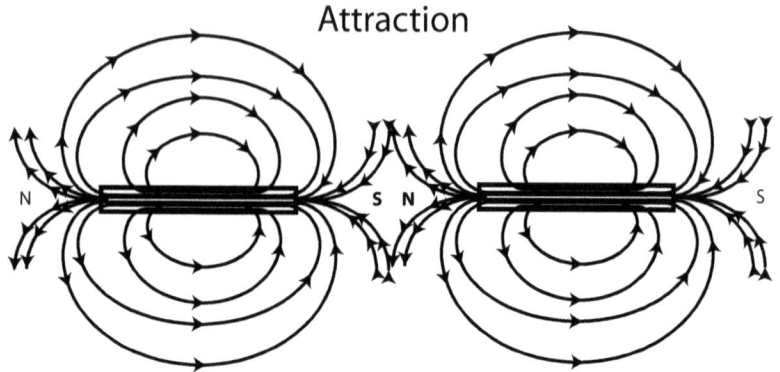

Figure 13.3. Lines of force demonstrating how unlike poles attract each other.

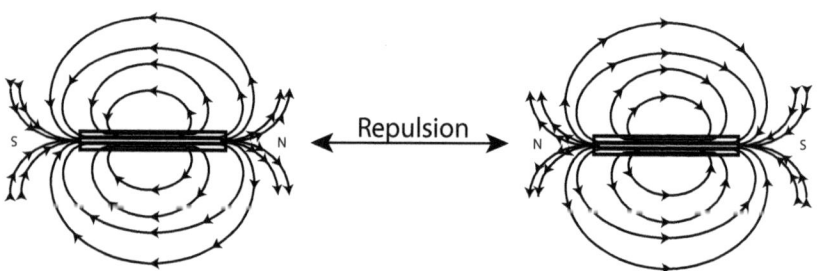

Figure 13.4. Lines of force demonstrating how like poles repel each other.

Electromagnets

Electromagnets are made by winding insulated copper wire around an iron rod. When electricity is passed through the wire, the iron rod becomes a magnet with a north and a south pole. Lines of force will be present and will behave like a naturally-occurring magnet. See figure 13.9.

Magnetic Fields

A **magnetic field** can be described by direction and magnitude. There are two parts to the phenomenon of magnetism. Magnetic charges are a product of electrical charges and charges attached to particles that are smaller than atoms. These particles cannot be broken down into any smaller particles. They are called **elementary particles**. **Quarks**[142] and **leptons**[143] are examples of elementary particles. There are different kinds of quarks and leptons. They are discussed in the last chapter.

142 There are six kinds of quarks that make up protons and neutrons.
143 There are six kinds of leptons. One we know as an electron.

Do not confuse elementary particles with atoms. Yes, atoms are the smallest particles of matter and cannot be broken down into smaller particles and still be that particular element. The key phrase is "cannot be broken down into smaller particles and *still be that particular element.*" We know today that atoms can be broken down into smaller particles, namely protons, neutrons and electrons, except for hydrogen-1 which has only one proton and one electron and a mass of 1 ($_1H^1$).

When charged particles are in motion, magnetism results. For example, the elementary particle we call an electron when in motion in a wire produces a magnetic field.

Effect of Electric Current
Electrolysis of Water

Direct current from a 12 volt battery can decompose water into hydrogen gas (H_2) and oxygen gas (O_2). This is ***electrolysis*** of water.

A water molecule contains two hydrogen atoms for each atom of oxygen. The oxygen atom has more electrons in its orbital than the hydrogen atoms, creating a polar molecule. It is like a tiny magnet. See figure 13.5.

The negative side of a water molecule is attracted to the positive end of another water molecule. Water molecules are attached to each other by weak hydrogen bonds. Boiling breaks these bonds, but cannot break the covalent bonds that hold hydrogen atoms to oxygen atoms. Elecrticity can break covalent bonds.

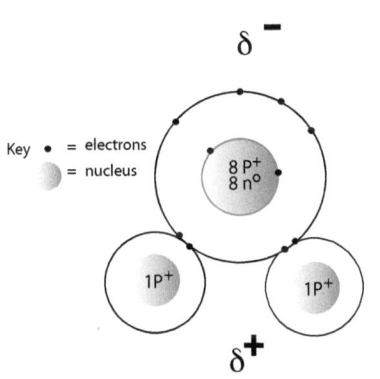

Figure 13.5. A water molecule. The area around the hydrogen atoms is designated positive (+) and the area around the oxygen atom is designated negative (−).

When an electric current is passed through water, covalent bonds that hold hydrogen atoms and oxygen atoms together are broken. Oxygen and hydrogen atoms are released. The free hydrogen and free oxygen atoms quickly unite for form molecular hydrogen (H_2) and molecular oxygen (O_2). The test tube containing a negative electrode will attract hydrogen gas and the positive electrode will attract oxygen gas molecules. This occurs because hydrogen molecules are more positively charged and oxygen molecules are more negatively charged. Opposites attract. See figure 13.6.

Figure 13.6 shows that there is twice as much hydrogen collected compared to oxygen. If a burning splint is placed in the test tube with hydrogen gas, the gas will burn rapidly. This is a characteristic of hydrogen gas. If a smoldering splint (glowing, but not burning with a flame) is placed in the test tube containing oxygen gas, the smoldering splint will burst into flame. This is a characteristic of oxygen gas.

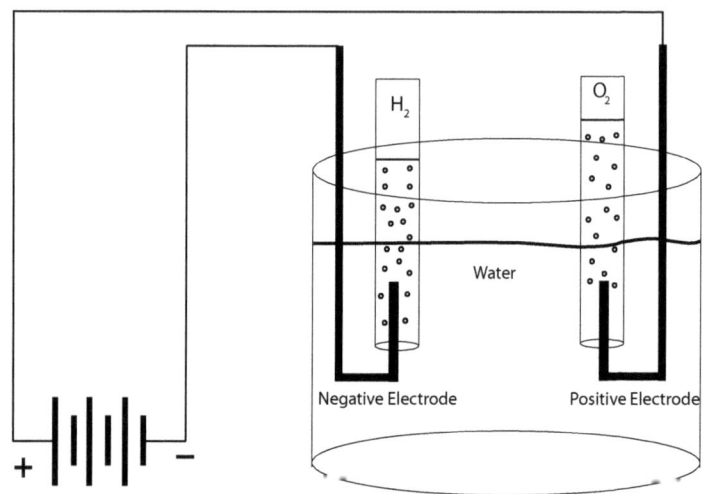

Figure 13.6. Electrolysis of water.

Based on the flame test for hydrogen and smoldering splint test for oxygen, it can be concluded that hydrogen and oxygen are the gasses released by electrolysis. Therefore, a molecule of water must have hydrogen and oxygen in its formula. Since the ratio of hydrogen gas collected to the amount of oxygen gas collected is 2:1, then there must be 2 hydrogen atoms for each oxygen atom in the formula for water. The formula for the water molecule was determined in this manner. See figure 13.7.

$$2H_2O \xrightarrow{\text{Electricity}} 2H_2 + O_2$$

Figure 13.7. A balanced chemical equation illustrating the electrolysis of water.

Incandescent Light Bulb

Thomas Edison was granted a patent for his improved electric light bulb in 1880. There had been a great deal research in this area, but Edison was granted his patent in 1880 for an improved, commercially-viable product. Edison continued to improve his light bulb and was able to have one last 1,200 hours using a carbon filament in a vacuum. Others discovered that filaments made of the element tungsten outlasted all other bulbs. In 1913, an American found that filling a glass bulb with an **inert gas**[144] made the filaments burn brighter and longer than in a vacuum alone.

The electric light bulb is a glass bulb with a tungsten filament through which current passes, making the filament glow or incandesce. The bulb is filled with argon or nitrogen, both inert gases.

Most of the energy used in an incandescent light bulb is given off as heat, as much as 90 percent. Heat wastes electricity. Incandescent light bulbs are a very inefficient way to light a space in the home or in the workplace.

Fluorescent Bulbs

Fluorescent light bulbs are about four to six times more efficient than incandescent light bulbs. Fluorescent light bulbs use electricity from a filament that gives off electrons that excite a vapor of atoms of mercury, which in turn make a phosphorescent coating on the inner surface of the tube glow.

Magnetic Effects of Electric Current

The Danish scientist, Christian Oersted (1777-1851) discovered that when a compass was placed near a wire carrying a current, the needle deflected and aligned itself perpendicularly to the wire. If the current was reversed, the direction of the compass needle was reversed. This was one of the early proofs for the existence of a magnetic field. See figure 13.8.

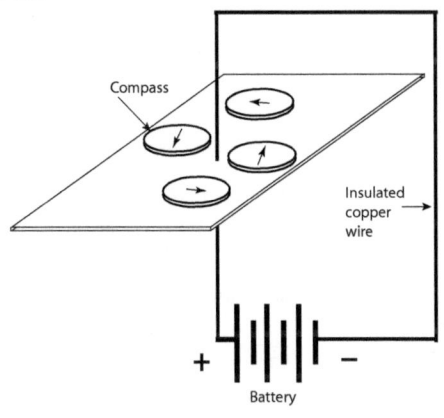

Figure 13.8. Four compasses surrounding a current carrying wire.

---

144 Inert gases are gases that will not combine with other substances.

Electromagnets

We can see from figure 13.8 that an electric current running through a conducting wire sets up a magnetic field. If the same current carrying wire is wound around a bar of iron, the loop of the magnetic field is set up around the iron core. See figure 13.9.

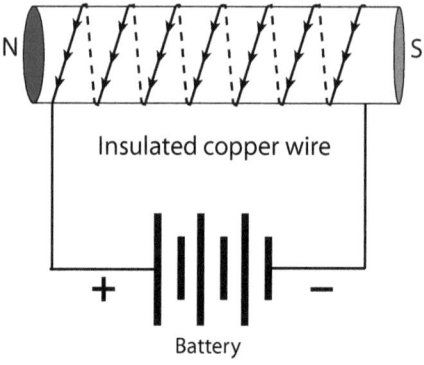

Figure 13.9. Electromagnet made by wrapping insulated copper wire around a soft iron core.

Direct Current

Direct current (DC) is a one way flow of electrons. Batteries were the first sources of direct current. Invented by Alessandro Volta (1745-1827) in 1800, the Voltaic pile was the first electricity-producing battery. It consisted of a disk of copper or silver and zinc separated by a disk soaked in salt water. See figure 13.10.

Today, direct current can be produced by solar cells and **dynamos**[145] as well. Direct current is different from alternating current (AC) which flows alternately back and forth in conducting wires.

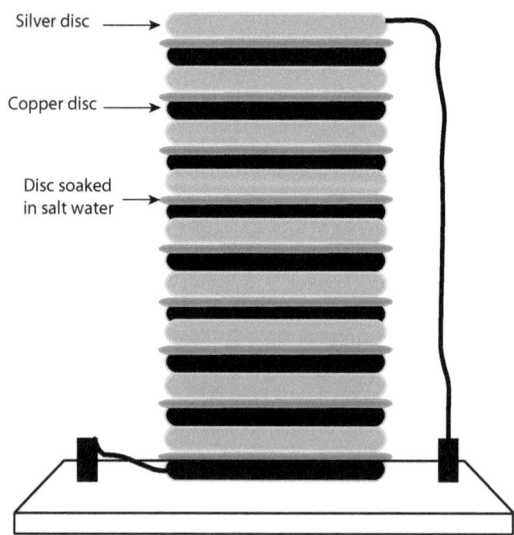

Figure 13.10. Voltaic pile.

145   Another name for an electrical generator that produces direct current.

Alternating Current

An electric charge that reverses its direction periodically is called *alternating current*. Current is a flow of electrons along a conductor of some sort, most often a copper wire. The ampere is the unit used to measure electric charge or current. A solution of charged particle (ions) can also carry an electric charge.

Direct current will be seen on an oscilloscope screen as a straight line, as opposed to alternating current which appears as a *sine* or *sinusoid* wave. A sine wave shows the reversal of electric current. See figure 13.11.

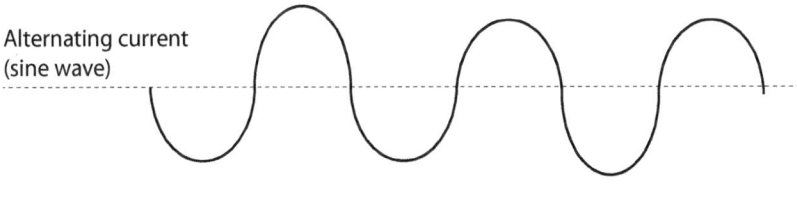

Figure 13.11. A sine or sinusoid wave versus direct current.

Generators

Generators are devices that use mechanical energy to produce electricity. Generators use the kinetic energy of steam, falling water, wind or internal combustion engines to turn a device that interacts with magnetic fields to produce electricity.

Motors

An *electric motor* is a device that converts electrical energy into mechanical energy. A DC motor consists of two principle parts: an armature and magnets. The armature is a winding of a single strand of insulated wire around a metal core. Current from a battery is sent to the armature to produce a magnetic field and the magnetic field of the armature interacts with the magnetic field of the magnets surrounding the armature. See figure 13.12.

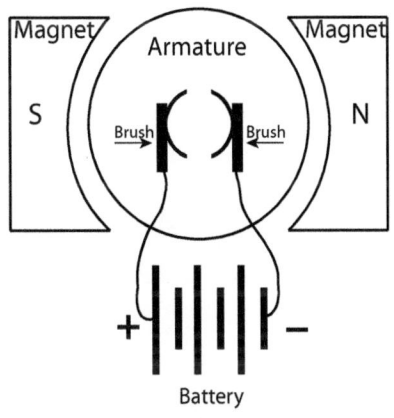

Figure 13.12. Diagram of a DC motor.

Transformers

A *transformer* is an electrical device that boosts or steps up an input current (the load) to a higher voltage or it can step the load down to a lower voltage. Electricity produced in power plant generators is too high to be useful to a consumer in the home. The voltage has to be stepped down first to a suitable voltage that can be distributed to homes and apartments, usually 600 volts. The 600 volts are carried along *distribution wires* suspended on utility **poles**[146] on the street in many neighborhoods. These are the service wires that carry a reasonably safe voltage to use for appliances. The voltage is stepped down further to 110 to 115 volts. See figure 13.13.

Sometimes step down transformers are mounted on a utility poles. The electricity from this pole-mounted step down transformer is what enters the residential dwelling. The point of entry for the customer is at a *service mast* or *weatherhead*. The weatherhead can be seen as a U shaped metal cap on the top of a plastic **conduit**[147] that begins on the meter pan and goes to the highest point of the house. It is done this way so that it cannot be touched. The electrical lines that enter the masthead from the utility poles are called the *service drop*.

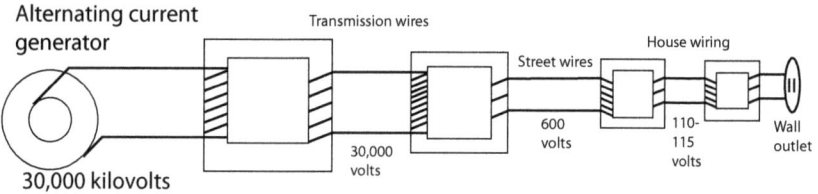

Figure 13.13. A series of step down transformers showing how voltage from a power generating station is brought down to a safe and usable voltage in the home.

---

146 In Manhattan, one of the five boroughs of New York City, all electrical transmission lines were removed from utility poles and buried beneath the streets. After the blizzard of 1888, it was determined that the complex network of wires was dangerous.
147 A tube through which electrical wires pass.

Electronics

Electronics is a branch of physics that deals with electrical circuits that control electron flow using transistors, diodes and integrated circuits and at one time, the commonly used vacuum tube. Civilian and commercial application of microwaves include cellphones, computers, microwave ovens, televisions, global positioning satellite systems (GPS) for automobile travel (see figure 13.15) and recreational boat navigation.

Military uses of microwaves includes spy satellites, GPS guidance for ships and airplanes, aircraft radar and fire control for naval weapons.

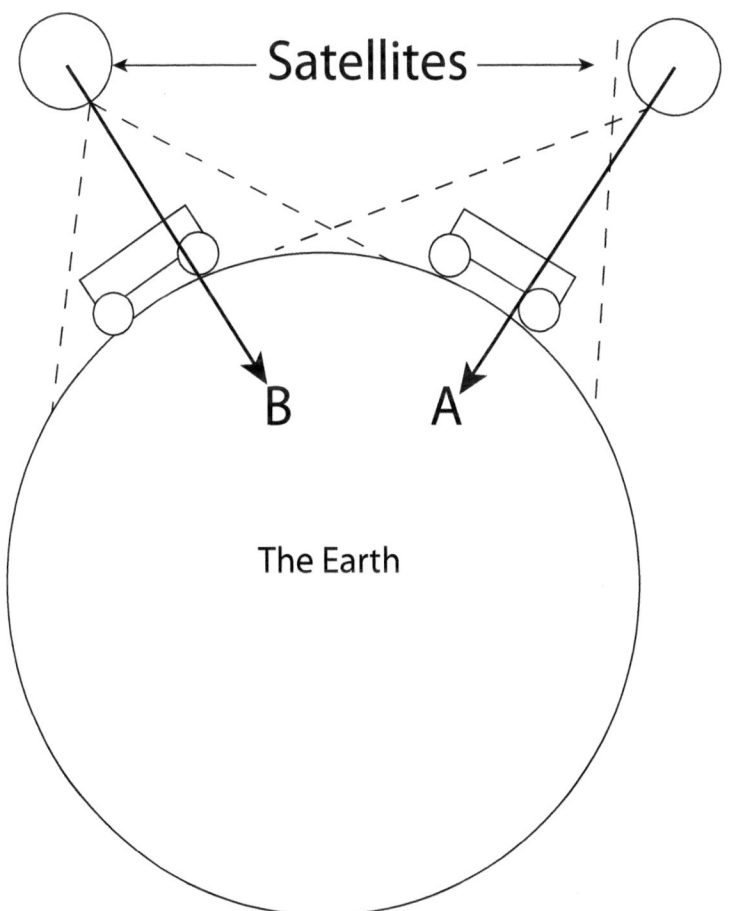

Figure 13.15. Global positioning satellites are on a "fixed" position above the Earth. The satellites have the same rotational speed as the Earth. As vehicles drive from one location to another they are detected by one satellite and then another.

Chapter 14 Radioactivity
Radioactive Elements and Radioactive Decay

Each element has at least one radioactive isotope that undergoes spontaneous decay. These isotopes give off ionizing radiation in the form of particles such as protons, neutrons, beta (ß-) particles (electrons), beta (ß+) particles (positive electrons), alpha (∂) particles and gamma (γ) rays. Atoms called *radionuclides* or nuclides for short, are produced as a result. See figure 14.1 and table 14.2.

Henri Becquerel discovered radioactivity in 1896. The French scientist wrapped a uranium compound in photographic film. The result was a blackening of the film due to the radioactive particles given off by the uranium. Soon after, other scientists discovered other radioactive elements. Marie Curie, a French-Polish physicist and chemist, discovered polonium and radium in 1898. Radioactive elements pose serious health threats, unknown at the time of their discovery. "Tonics" containing radium were prohibited after it became obvious that people were dying from drinking them.

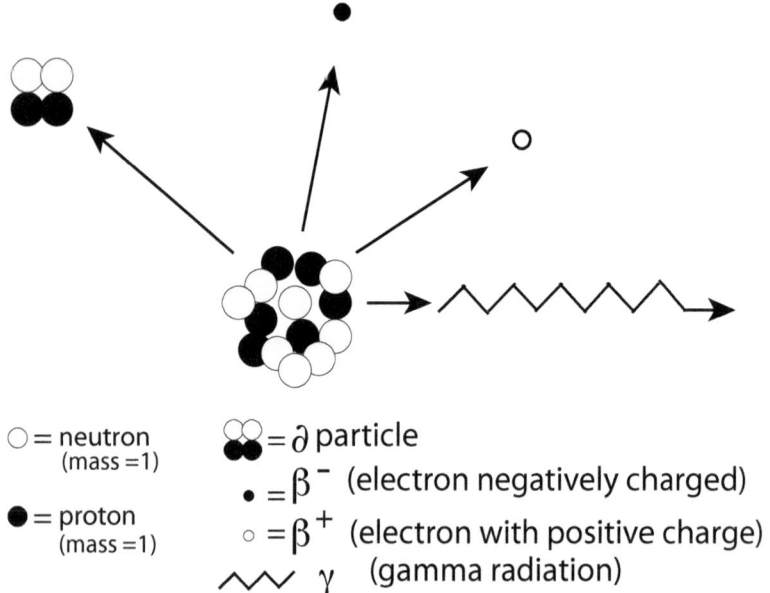

○ = neutron (mass =1)
● = proton (mass =1)
⚫⚫ = ∂ particle
• = β^- (electron negatively charged)
○ = β^+ (electron with positive charge)
∿ γ (gamma radiation)

Figure 14.1. General representation of spontaneous nuclear decay showing some of the major particles and radiation waves that may be given off.

Nuclides

A *nuclide* is an atomic nucleus described by a specific number of protons referred to as the *Z number*, a specific number of neutrons, the *N number* and the *A number*, its mass. The energy state of the nucleus is also part of the description. The term "nuclide" was coined in 1947 to focus on the atom's nucleus and disregard electrons and their arrangement around the atom's nucleus. Each nucleus that has a different Z number is considered a species.

If two atomic nuclei have the same Z, but different N, then one is the isotope of the other. See table 14.2. Hydrogen has three nuclide species that have the same Z number, but different N numbers and consequently different mass numbers. Each is an isotope of the other. The three **isotopes**[148] are nuclides because the Z number is fixed.

Isotones are species of nuclides that have the same number of neutrons. For example, carbon has two nuclides that have the same number of neutrons. This is determined by subtracting Z from A for both nuclides. Carbon's A is 13 and its Z is 6 (13 - 6 = 7). Nitrogen's A equals 14 and its Z equals 7 (14 - 7 = 7).

Isobars are nuclides that have equal mass numbers. Oxygen, nitrogen and fluorine all have the same mass number. See figure 14.2.

Isotopes have equal isotope numbers, but different mass numbers

```
12 ←— Mass (A) number —→ 13
C                          C
6  ←— Proton (Z) number —→ 6
Example 1.                 Example 2.
```

Isotones have equal neutron numbers

```
13              14
C               N
6               7
```

Isobars have equal mass numbers

```
17        17      17
O         N       F
8         7       9
```

Figure 14.2. Examples of isotopes, isotones and isobars.

---

148   Chemical isotopes have the same number of protons but different numbers of neutrons.

Half-Life and Radioactive Decay

Half-life ($T_{1/2}$) is a way of describing the rate at which a radioactive element decays into a simpler element. In the first half-life period, half of the element changes into another element. Half of the original sample of the **radionuclide** is no longer present. In the second half-life period, half of the half that did not change in the first half-life period changes into another element. Now one-quarter of the original sample is left. After the third half-life period, one-eighth of the original sample of the radionuclide remains and seven-eighths is another element. See table 14.1.

Table 14.1. Half-lives of a few radionuclides.

Element	Symbol	Half-life
Uranium	$_{92}U^{238}$	4.5 billion years
Radium	$_{88}Ra^{226}$	1,620 years
Polonium 210	$_{84}Po^{210}$	138 days
Radon	$_{86}Rn^{222}$	3.8 days
Polonium 214	$_{84}Po^{214}$	1.64 x 10^{-4} seconds

There are about 255 nuclides that have never been detected to display spontaneous radioactive decay. They are stable nuclides. There are nuclides that are naturally unstable (see table 14.2), that is, they undergo spontaneous nuclear decay (see figure 14.1) and are called natural radionuclides. **Natural radionuclides** may be classified into three categories. First, there are the primordial[149] naturally-occurring radionuclides. These natural radionuclides were formed in stars before the formation of the Earth's solar system. They are called primordial because they are left over from the process of nucleosynthesis that occurred at that time. Uranium 238 (^{238}U) is one example. Its half life is 4.5 x10^9 years. Uranium 238 is still an abundant element, but U-235 is over 130 more times more rare. Its $T_{1/2}$ is 0.7x10^9.

A second category, called **daughter nuclides**, form from the decay of naturally occurring radionuclides such as uranium or thorium. Radium is one of these and has a relatively short half life. Its $T_{1/2}$ is 1,602 years. Francium is another example of a daughter nuclide.

A third group is called **cosmogenic nuclides**. They are formed continuously by naturally occurring nuclear reactions that take place in the nucleus of a single atom. The Earth is under constant bombardment with protons, atomic nuclei, or electrons that are called cosmic rays. These particles penetrate the Earth's surface and interact with atoms and

149   Primordial: a beginning, something that gave rise to something else, original.

cause the formation of carbon-14 (C-14), berylium-10 (Be-10), chlorine 36 (Cl-36). The amount of these radionuclides is helpful in dating rocks that range from 1,000 to 30,000 years old.

Table 14.2. A partial list of natural radionuclides.

Element	Nuclide	Atomic Mass Number (A)	Proton Number (Z)	Neutron Number (N)
Hydrogen	$^3H_1$	1	1	2
Helium	$^5He_2$	5	2	3
Lithium	$^8Li_3$	8	3	5
Carbon	$^{14}C_6$	14	6	8
Nirtogen	$^{16}N_7$	16	7	9
Potassium	$^{40}K_{19}$	40	19	21

Carbon-14 Dating

Depending on the half-life of a radioactive nuclide, the nuclide can be used to date a fossil. *Half-life* is the amount of time required for a sample of a radioactive element to decay into another element. There are three isotopes of carbon in nature. Carbon-12 makes up about 99% of the carbon on Earth. Carbon-13 makes up about about 1% and carbon-14 is only about 1 part per trillion (0.0000000001%), but measurable.

Cosmogenic C-14 is formed in the atmosphere by neutron bombardment from cosmic rays and has a half-life of about 5,730 years. Living things will have a constant amount of C-14 in their bodies while they are alive. When they die, however, the amount of C-14 they accumulated in life remains at that level. The age of the fossil can be calculated by determining how many half-lives C-14 has undergone. This is done by comparing the ratios of C-14 to C-12. For example, a sample has been found to contain C-14 that has undergone two half-lives. Since the half-life of C-14 is 5,370 years, then the age of the sample is 10,740 years. Analysis of the sample is done to determine how much C-14 is left.

Nuclear Fission

Nuclear fission is splitting one large atomic nucleus into two smaller nuclei with the release of energy. See figure 14.3. A single neutron can produce a nuclear chain reaction. If the nucleus of a U-235 is split, each half produces more than one neutron. Each of these splits other U-235 nuclei. See figure 14.4.

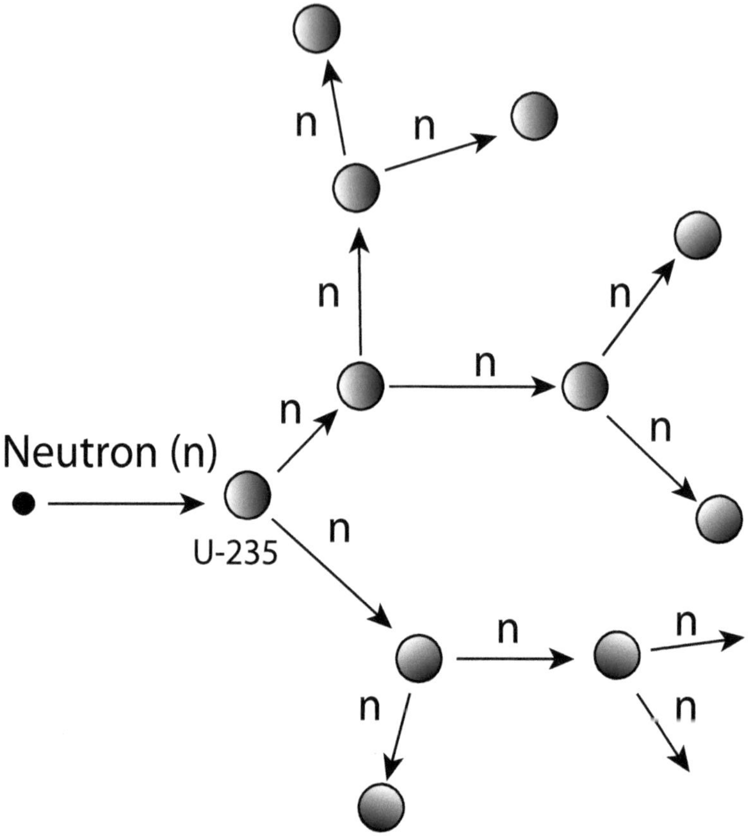

Figure 14.4. Chain reaction produced by nuclear fission.

Nuclear power plants use controlled nuclear fission reactions to produce thermal energy that boils water to produce steam. The steam spins turbines attached to electrical generators to produce electricity.

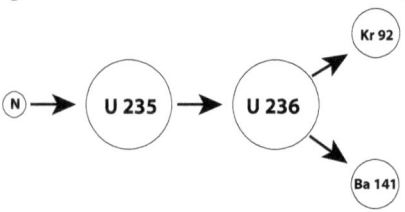

Figure 14.3. Nuclear fission.

A fission reaction (see figure 14.3) can produce huge amounts of thermal energy that, in an uncontrolled chain reaction, can destroy cities. These are atomic or nuclear weapons. Atomic weapons have been used twice in the history of warfare. Albert Einstein predicted the release of huge amounts of energy from a small mass using in his equation:

$$E = mc^2$$

E represents energy, **m** stands for mass and **c** represents the speed of light (186,000 mi/s or 300,000 k/s). So, if a little less than 1 kg of U-235 is multiplied by 186,000, the equivalent of 16,000 tons of TNT in explosive energy is released. This is about the amount of mass that fissioned in the atomic bomb that was dropped on Hiroshima, Japan in August, 1945. The results were so devastating that countries are not willing to use them again to resolve disputes.

Nuclear Fusion

Nuclear fusion is the joining of two small atomic nuclei to form one nucleus with the release of tremendous energy, even more than in fission reaction. Stars such as the Sun are powered by this process. See figure 14.5.

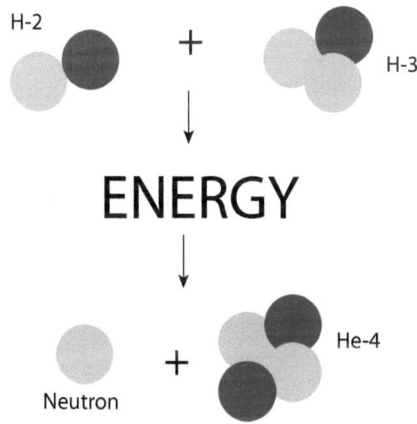

Figure 14.5. Representation of a fusion reaction.

Detecting Radioactivity and Radiation Counters

Ionizing radiation safety is a primary concern for some research scientists, health care workers in hospitals, nuclear power plant workers, emergency response teams and those dealing with hazardous materials. Ionizing radiation detectors detect particles possessing high energy that result from nuclear decay and cosmic radiation.

Any worker that is a risk of exposure to ionizing radiation must wear dosimeter badge. The dosimeter consists of photographic film in a holder. Badges are regularly developed after a specific period of time to determine if the worker has been exposed to higher than allowable levels of radiation.

Some health care workers, such as radiologists, have the greatest exposure to radiation. Most nurses do not. Some patients undergo procedures that involve nuclear medicine in one form or another such as bone scans, gallium scans and **PET scans.**[150] Radioactive tracers are used in such small amounts they pose virtually no hazard to a patient or a nurse.

---

150 Computerized tomography (CT), magnetic resonance imaging (MRI) and ultrasound scans do not involve radioactive materials.

Unified Physics

There has been an ongoing effort to unify quantum mechanics (quantum physics) with Albert Einstein's theory of relativity. Quantum means extremely small. As a matter of fact, a quantum is the smallest amount of any kind of matter that can interact with anything else. If a quantum is the smallest amount of a physical entity, then it can be measured. A number can be attached to this physical quantity. In other words, it can be interpreted as a quantity, a "quantum." Quanta are among the smallest quantities known to exist. The universe is thought to have been created from a quantum of matter of immense mass and density that "exploded" about 14 billion years ago. This is termed the big bang theory of the creation of the universe.

Early thoughts concerning quantum mechanics were developed in the first decade of the twentieth century. Around that time, great progress was made in atomic theory and the theory of the corpuscular nature of light. Atomic theory became accepted as fact, but the theory of the corpuscular nature of light was, in reality, quantum theory dealing with matter and electromagnetic radiation. Leading scientists emphasized measurement in quantum mechanics. The twentieth century saw additional studies of quanta: quantum chemistry, quantum electronics, quantum optics, and quantum information science to name a few.

www.ingramcontent.com/pod-product-compliance
Lightning Source LLC
Chambersburg PA
CBHW040806200526
45159CB00022B/33